WOMEN IN AGRICULTURE

WOMEN IN AGRICULTURE
A Socio-Economic Analysis

SHASHI KANTA VARMA

CONCEPT PUBLISHING COMPANY, NEW DELHI 110059

ISBN 81-7022-363-6

First Published 1992
Reprinted 2001

Printed and Published by
Ashok Kumar Mittal
Concept Publishing Company
A/15-16, Commercial Block, Mohan Garden
New Delhi-110059 (India)
Phones: 5648039, 5649024
Fax: 091–(11)–5648053
E-mail: publishing@conceptpub.com

DEDICATED
TO MY
BELOVED GRANDFATHER

ACKNOWLEDGEMENT

I take the privilege to express my esteem and profound sense of gratitude to Dr. B.P. Sinha, Principal Scientist and Head of the Division of Agricultural Extension, IARI, New Delhi, for his learned counsel, sagacious guidance, kind encouragement and constructive criticism all along the investigation and organisation of this study. His sympathy and generosity have been much beyond his formal obligation as the chairman of my advisory committee, for which I am deeply indebted to him.

I express my profound sense of gratitude to Dr. Y.P. Singh, Professor, Division of Agricultural Extension, IARI, New Delhi, one of the distinguished member of my advisory committee for his immense interest, constant and ungrudging help and affectionate behaviour.

I am highly indebted to Dr. S.S. Pillai, Ex-Joint Director, computer section, IASRI, New Delhi for his continuous help and timely suggestions during the course of this study. I am also grateful to Shri Mahesh Kumar and Shri M.L. Chaudhry, programmer, IASRI, New Delhi for their kind cooperation in analysing the research data.

I record my sincere appreciation and greatfulness to the members of my advisory committee, Dr. C.M. Bhat, Scientist, Division of Agricultural Extension, IARI, New Delhi and Dr. Randhir Singh, Professor, IASRI, New Delhi for their continuous help and valuable suggestions during the course of this study.

I am highly indebted to Dr. P.N. Mathur, ADG (Extension), ICAR, New Delhi and Dr. R.P. Singh, Senior Training Specialist, Division of Agricultural Extension, New Delhi for their wise counselling and guidance throughout the present endeavour.

I am greatful to my learned teachers, Dr. Rekha Bhagat, Dr. Anita Jhamtani, Dr. K. Vijay Raghvan, Dr. A. Ramesh Babu and Shri B.N. Chattopadhyay and Dr. D.U.M. Rao for having broadened my cognition-span.

The instrument of acknowledgement would remain out of tune if I do not express my thanks to the respondents who extended their unreserved cooperation during data collection.

I extend my hearty thanks to Dr. V.N. Tripathi, Scientist, NDRI, Karnal and his family for their love and encouragement, which have been a fountain of great inspiration to me.

I take this opportunity to express my appreciation to silent support extended by Mrs. B.P. Sinha during the writing of the manuscript.

I express my deep feelings of indebtedness to Dr. T.R. Verma, Professor, Department of H.Sc. Ext. Edu., HAU, Hisar and her family for their consistent help and encouragement throughout this study.

Every researcher owes a great deal to her friends, and I am no exception. I cherish my close association over the years with my classmates Mrs. Reeta Bansal, Mrs. K. Vijaya, Mr. Nirmal Kumar, Mr. Didar Singh, Mr. O.P. Singh, Dr. S.N. Ojha, Mr. R.A. Dey, Mr. N. Kishore Kumar, Mr. Ravindra Padaria, Mr. Amarnath Gupta, Vijayalakshmi, Jasdeep Chatrath and Anju Gupta.

With a risk of serious omission, I wish to sincerely acknowledge the help of Mr. Dipankar Kar, Mrs. Reeta Kar, Dr. M.A. Akbar and Anjani Kumar.

I extend my thanks to Mr. Krishan Kumar Sharma and Mr. Bhim Sain Singal for taking hard pains in meliculaus typing of this manuscript.

I am deeply indebted to Haryana Agricultural University, Hisar for having provided this excellent learning opportunity by sponsoring me at Indian Agricultural Research Institute, New Delhi for doctoral programme. Also, I record my thanks to ICAR for offering me a Senior Fellowship for my Ph.D. programme.

I owe an encompassing debt to my parent's filial affection, encouragement and blessings without which this study would not have been completed. In the same vein, I appreciate the over-willing help of my sister Manju and brothers Sanjay and Rakesh during this study.

Finally, I owe a great deal to my husband Dr. Ajay Kumar Varma for his unfathomable love, encouragement and inexhaustible help without which this study would have remained a chimera.

SHASHI KANTA VARMA

CONTENTS

LIST OF TABLES

LIST OF FIGURES

1

INTRODUCTION

"In order to awaken the people it is the woman who have to be awakened. Once she is on move, the family moves, the village moves, the nation moves".

— *PANDIT JAWAHARLAL NEHRU*

Women are almost half of the adult population. They constitute one-third of labour force but consume two-thirds of the world's working hours and yet earn only one tenth of the income and own only one per cent of world property (United Nations, 1975).

An analysis of International Labour Organisation, Food and Agriculture Organisation and National Population Census data was carried out by Dixon (1982) which reveal that women constituted 38 per cent of agricultural labour force in developing countries. For 19 countries of South and South East Asia, Dixon (1982) estimated that 45.3 per cent of agriculture labour force consist of women. According to him this figure in fact is an underestimate because of defects in data collection which include under counting of the contribution of unpaid family labour, underestimate of seasonality of women labour and self-reporting bias of the interviewees who are by and large menfolk.

The real issue therefore, is more serious, despite their involvement in agriculture work in such a long magnitude they have not been actively involved in the main stream of development and there is hardly any appreciation and recognition of their extensive contribution. By and large, they have remained as "invisible workers".

The prosperity and growth of a nation depend on the status and development of its women as they not only constitute nearly half of its population, but also influence growth of the remaining half of the population. The crucial role of women in agriculture, allied occupations and household activities have, however, been underestimated and undervalued.

The multiple roles played and the productive inputs made by women in terms of work hours contributed or equivalent income generated in the family are neither attended nor recorded.

The value of unpaid household work constitute 25-39 per cent of the total gross national product (GNP) in developing countries (International Labour Organisation report, 1975).

According to the 1981 census, in India, there are 32 crore women out of the total population of 65.8 crore (48.6%). The primary sector consist of farming, livestock, forestry, fishery, plantation and orchards etc. provides work to about 11 per cent of the total women population and about 87 per cent of the total women work force. Women constitute 46.5 per cent of total population in the State of Haryana. About one-third of the women population of the State is engaged in agricultural activities which is three times higher than the comparative national figure of women work force (33% as against 11%). Women do more field-work than men is a matter of common knowledge particularly in the State of Haryana (Acharya and Patkor, 1983; Ghosh and Mukhopadhyay, 1984; and Mies, 1984).

Women play a significant and crucial role in agricultural development and allied fields including crop production, livestock production, horticulture, post-harvest operations, agro/social forestry, fishing etc. The nature and extent of women's involvement in agriculture, vary greatly from region-to-region, Even, within a region, their involvement varies widely among different ecological sub-zones, farming systems, caste, classes and socio-economic status of families etc. (Swaminathan, 1985). Regardless of these variations, there is hardly any activity in agricultural production except ploughing in which women are not actively involved.

ILO (1979) reported that if a total range of women's actual economic activities were to be taken into account, all over India, women work for very long hours in market and non-market economy, and more so in the rural India. It is the nature and type of their work and the long work hours which raise fundamental economic and social issues. These issues are related to the stage of technological development of an economy as well as the existing division of labour between men and women in the society.

Srivastava (1985) further reported that women irrespective of land status of the family provide 14 to 18 hours of productive physical labour in different chores. The energy spent by them in performing these tasks is more than it is physically feasible for them to spend particularly in a below subsistence level of living. This draws our attention to the plight and drudgery of rural women. The State of Haryana where this study was conducted is no exception.

Drudgery is generally conceived as physical and mental strain, fatigue, monotony and hardship experienced by human beings. While all of these results show decline in living and working conditions affecting men and women alike, the plight of the poor and the weak in this regard is alarming as they continue to be constrained by illiteracy, malnutrition and unemployment.

One of the ways to substantially reduce drudgery is to mechanise agricultural operations. Castillo (1977) made a startling revelation that even mechanisation in agriculture involve mainly those operations which are largely performed by men, while women's work in both on and off the farm (including household tasks) has been predominantly manual. She further added that much of such technological innovations had absorbed male labour and disemployed female labour. This generalisation appears to apply to multiple agricultural tasks (such as weeding, harvesting and carrying operations), food processing and a wide range of construction activities. This view was supported with specific instance by Banerjee (1985) who said that with introduction of weedicides and pesticides, a large number of poor women lost their meagre source of income. According to him it is tautology that technological change and improvement in any sphere, trade and craft gives it status, increases economic return and make it the purview of men. Women meanwhile continue to cluster in the least skilled, most monotonous, most indispensable and low paid tasks in the production process.

Women's participation in agriculture is adversely affected by modern method of cultivation, which results in gradual displacement of women and shrinking of their activities. Similar conclusions were arrived at in the studies of Whyte (1978) in Java, Kabir and Chen (1978) in Bangladesh, Whyte and Whyte (1978) and Hanger and Maris (1973) in Kenya, Palmer (1977) and Nancy and Bay (1970) in Africa and Dey (1985) in India.

Modernisation of agriculture has provided women with better income earning opportunities in some areas but has displaced them from their traditional roles in others. Such a negative impact on employment could cause much distress particularly to families where the total family income is small and hence needs to be supplemented by the earnings of both husband and wife. All these aspects require a close examination.

Since 1970, a global concern for women's emancipation all spheres has been expressed. In India such efforts have also been made by different Ministries/Departments and Non-Government Organisations, but in a vast country like this, the impact of such efforts have hardly been enough. In case of agriculture where involvement of women is far more prominent, the situation is no better, In fact, the agricultural development system has not yet taken farm women population seriously in the main-stream of its development processes. Therefore, it is high time that this issue is discussed and deliberated at the national and regional levels in order to sensitize people about its potentiality, prospects and obstacles and charting the course of actions devoted to bringing women as active partners in research and development (R & D) with equal rights and privileges.

"During the UN Decade of women, which ended in the year 1985, it became apparent that official statistics do not reveal fully the actual status and role of women in agriculture. Such data are necessary for economic analysis and policy decisions in agricultural planning, food security, alleviation of rural poverty, rural development and agrarian reform. FAO has already launched a special data collection programme with a component of statistics on women in agriculture as follow up of the world conference on Agrarian Reform and Rural Development Programme of Action".

Like many other countries, India lacks adequate quantitative data on women in agriculture. Farm women constitute so significant a part of the working women population in our country that it necessitates a fuller understanding of their status and role not only as they now are but as they may be developed in the future. Such understanding will mirror, at least to some extent, allocation of their time to all types of activities within and outside the home, the magnitude of their contribution to agricultural development in a broder sense of the term, their potential, constraints and employment calls for an improvement in skills and technologies for increasing their productivity. Quite a few but sporadic and lop-sided studies on women in agriculture are, no doubt, available. They have been conducted under various limitations and so most of these studies do not reflect a clear and complete all

India picture as to the nature and extent of farm women's participation in a wide range of household and non-household activities. As such, much additional research is needed before a comprehensive grasp of their distinctive role as home makers and farm workers and the problems they encounter in effectively discharging their responsibilities, can be had.

In view of the foregone expositions, the major issues which the present investigation proposes to address itself are the following:

1. The multiple role of women in agriculture, allied occupations and household activities are generally, underestimated and undervalued. By and large, they have remained as "invisible hands".
2. The nature and extent of women's involvement in agricultural and non-agricultural activities vary greatly from region-to-region and even within a region, it varies widely among different farming system and socio-economic status group of families.
3. Women toil hard in labour intensive jobs but have little access to the knowledge of modern farm technology.
4. Women's involvement in agricultural tasks by and large is a source of heavy burden of drudgery on them. Drudgery ridden agricultural tasks are assigned to women probably because of male dominated social order.
5. Whenever technological innovation(s) to reduce drudgery on a particular task, is available, it results in displacement of women from the tasks and sometimes rendering them unemployed (partially or fully).

These assumptions are conjectural in nature and seem to have found their expression only in opinion or subjective generalisations. There had been on attempt to test these assumptions empirically. The present investigation was, therefore, designed to test the tenability of these assumptions through empirical analysis of data with the following specific objectives:

1. To study the nature and extent of involvement of rural women and men of various socio-economic strata in agriculture and home activities in terms of operations, time spent, responsibility and accountability.

2. To study the extent of exposure of modern crop technology to farm women.
3. To prepare a comparative drudgery index for agricultural operations generally taken up by/assigned to women and men in the community/locality.
4. To study the impact of mechanisation on farm women especially in terms of employment, displacement, work load and income.
5. To study the extent of contribution of women in cultivation of major crops.
6. To study the role of women in decision-making in farming, home and other activities.
7. To study the utilisation pattern of savings in the family.

Scope

The present study with holistic approach will generate useful information on various aspects of involvement of women in different on farm and off farm activities, decision-making, economic contribution and drudgery undertaken by them. It is appropriate to quote Abraham Lincon's dictum "If you could but know where you are now and where you ought to go, you could better judge what to do and how to do it". This study, therefore, is of great significance in creating data base for a fuller understanding of the existing situation and also for the course of action to be taken in future.

Providing an insight into a typical agrarian profile of farm women, this study will suggest several far reaching implications to the planners, policy makers, administrators, scientists and change agents for transforming an agricultural based rural society.

Limitation of the Study

This study suffers from all the limitations of a student's research project. Time and money being the main constraints, the study was confined to only a few villages of a State with a total sample of only 100 (couples, husband and wife of the family) respondents. However, proper care and considerate thought have been exercised in making the study as empirical systematic as possible.

2

THEORETICAL ORIENTATION

Nearly half the available human resource in India, like many other countries, is women. Of the total female population of 318.25 million in India (1981) which is 48.35 per cent of the total human population, 245.12 million (77%) comes from rural areas whose major occupation is agriculture and agricultural based enterprises.

According to Swaminathan (1985) "some historians believe that it was women who first domesticated crop plants and thereby initiated the art and science of farming. Women have played and continue to play a key-role in the conservation of basic life support systems such as land, water, flora and fauna. Without the total intellectual and physical participation of women, it will not be possible to popularise alternative system of land management to shifting cultivation, arrest gene and soil erosion and promote the care of the soil and the health of economic plants and farm animals".

The role of Indian women, as it has evolved, been experienced and understood over 4,000 years, has been interwined with the history of the country which is primarily one of repeated impositions of an alien ethos on its culture necessitating a frequent restructuring of social systems and consequently, individual identity. Like the country, the role of Indian women is full of diversity and contradictions. Indian history is experienced from its glorious past to its degenerating poverty, from its spiritual ascendancy to its revelling exhibitionism, from freedom to captivity and from violent outbursts to tranquil moments of peace. Similarly, the role of Indian women has ranged from that of a *deity* to that of a *devdasi*, from the pure to the vulgar, from being supreme to being downtrodden, and as innumerable manifestations of virtue or vice.

The role of Indian women has undergone dramatic and drastic changes from era-to-era, while within the era themselves there have existed simultaneous contradictions. Furthermore, it has varied from caste-to-caste and with the various socio-cultural and economic strata of society. It is as panoramic as Indian history.

In ancient India, women had no less than divine status of Laxmi, Durga and Saraswati, – the Goddesses of economic prosperity, power and wisdom, respectively. A very popular ancient scripture declares that Gods live where women are respected.*

The roles women play in society and the images we have of them have developed not simply from the exigencies of biology and social situations but are rather deeply rooted in myths, legends and the religious rites and rituals of our culture. Philosophers, sages and theologians have at one time or the other paid attention to women and their socio-personal status. But unfortunately this did not flow smoothly in our male dominated society. In a society where men have controlled knowledge and have interpreted the classical texts, it is not surprising, as Ruth (1980) says that women have lost the power of explaining and defining for themselves the realities of their own selves. In a patriarchal culture, what we get is a masculinist definition of ideals and images of women. These ideals and visions are not women's creations. They are not born out of their own experiences.

The psychological perspective views her (women) as passive, masochistic and considered mandatory for fulfilment and identity formation. Kakar (1988) in his paper, 'Feminine Identity in India' and Parikh and Garg (1989) in their book, 'Indian Women: An inner Dialogue' look at women's development and transition through the various stages of her cycle from a psychoanalytic perspective. A variety of social agents prepare her for motherhood, a role in which her identity finds completion. The inspiration of ideal womanhood comes from the mythical figure of *Sita*. For both men and women in Hindu society, the ideal woman is personified by *Sita*, the quintessence of wifely devotion, the heroine of epic Ramayana. Her unique standing in the minds of most Hindus, regardless of region, caste, social class, age, sex, education or modernisation, testifies to the power and pervasiveness of the traditional ideals of womanhood.

The ideal of womanhood incorporated by *Sita* is one of chastity, purity, gentle tenderness and a singular faithfulness which cannot be destroyed or disturbed by her husband's rejections, slights or thoughtlessness. This is the ideal, purveyed over and over again, in

* Manusmriti.

numberless myths and legends, through which the Hindu community has tried to mould the character and personality of its female members.

The concept of the female in Hinduism presents an important duality; on the one hand, the woman is fertile, benevolent – the bestower; on the other she is aggressive, malevolent – the destroyer. Two facets of femaleness relate to this duality, and perhaps provide a cultural logic for it. The female is first of all *Sakti* (energy/power), the energising principle of the universe. The female is also *Prakriti* (Nature) – the undifferentiated matter of the universe.

Understanding the dual character of the Hindu female's essential nature (her *Sakti* and *Prakriti*), provides a backdrop for understanding the rules and role models for women in South Asia. A central theme of the norms and guidelines for proper female behaviour, especially in the male dominated classical literature, is that men must control women and their power.

Mythology, written and oral, in Sanskrit and in the vernaculars, provides many examples of female behaviour and its consequences thus setting up explicit role models for the Hindu women. Folklore yields yet other beliefs about female behaviour. Finally, social organisation and structure mesh with, allow for, and reinforce these beliefs about the proper conduct of women (Wadley, 1976).

However, a multitude of derogatory attributes has been ascribed to women in post-Vedic period. According to Manu, a woman must never be independent. In childhood, she must be subject to her father, in youth to her husband, and when her lord is dead, to her sons. In this context, the role of women has been visualised as passive one, which unfortunately has guided the course of action in the past leading to their present status of socio-economic backwardness in general and neglect in the streams of development and scientific arena in particular. The role of women has to be visualised as described in the words of Rabindranath Tagore in Chitranganda. "The heroine speaks to Arjuna : I am no goddess to be worshipped, nor yet the subject of common pity to be brushed aside like a moth with indifference. If you desire to keep me by your side, in the path of danger and daring, if you allow me to share the great duties of your life, then you will know my true self."

In many ways, the role of women in Indian society remained unaltered for about a thousand years. It was only under the impact of British rule and the second wave of christianity that Indian society began resisting its decadence. In the late nineteenth century, with the establishment of the Indian National Congress (1885), a more active movement to lead women into the modern era began. Since then, women have been able to take up careers as teachers, doctors, lawyers, political leaders and social workers. However, the economic and other freedom available to career women, have not been able to change the traditional orientation and attitudes towards them.

Even at the time of rigid orthodoxies, there were women who transcended the narrow boundaries of the prescriptive roles. They had the courage to define and design new roles. In the process, they were maligned, abused and sometimes destroyed, but they suffered this social response with conviction and dignity.

Over the centuries, then, like many other phenomena, the existence of women in Indian society continued to experience flux and transition. And with every change, women were pushed further in the direction of becoming objects. Occasionally, there were times when the doors of captivity were opened, the burden of oppression decreased and women could walk towards new horizons. Yet, in their journey across the forbidden territories of society, the ghost of stigma continued to accompany them. These journey required courage, intense commitment, relentless struggle and a lot of lonely moments of pain, anguish and tears.

The various technological transition followed by major change in the economic processes and structure of the country played a major role in defining the socio-psychological status of both men and women in India. Changes in technology are the product of man's efforts to enhance his physical and psychological well-being and survival. The development and introduction of technology in a society lets loose economic and other forces which compel man to change and adapt his interpersonal behaviour to the new environment. This obviously leads to new definitions of role and status.

A comprehensive view of evolution of technology in man's life space and the resulting and continuing changes in social codes and forms of work, suggest that four major transitions have taken place in the last 4,000 years of known history. Each transformation has brought about changes in the definition, concept, nature and meaning of work, and the role of men and women in the social system. With each transition, women have had to contend with diminishing horizons, changing role definitions and societal values, and a reduction in the degree of freedom allowed them within the frame-work of the social design. Broadly speaking, the four periods are : (i) Foraging and hunting culture; (ii) Cattle culture; (iii) Agrarian culture; (iv) Industrial culture.

An agricultural society has its own distinctive symbiotic relationship with nature. Since the time of neolithic agriculture, this distinctiveness has lain in the central role of women in society and culture. It was she who primarily involved in 'gentling, nurturing and breeding'; it was her capacity for tenderness and love which gave the earliest agricultural settlements of man, their touch of security, receptivity, enclosure, nurture and it was she who made fully possible the growth of civilisation (Mumford, 1961). In the Indian context, the transition into the agrarian culture is reflected in the Hindu epic, Mahabharata. Lord Krishna represents the cattle culture and his brother, Balram, often referred to as Haldhar (holder of the plough), represents the agrarian culture.

The stabilisation of the agrarian pattern of life led to the evolution of a more complex social structure. The community was transformed into a society, where issues of stratification, hierarchy, governance, maintenance and interdependence of different occupations, roles and sexes emerged. The impact of the agrarian society was more pronounced on women than on men. The entire scope of their role, the quality of their relationship with others, the very meaning of their existence everything was redefined. They were pushed into being appendages of men, with no independent status of their own. Men became the be-all and end-all of their existence.

Agrarian Indian society eventually relegated women to the exclusive role of the housewife. To whichever community or caste she belonged, her dominant space was at home. The role of man was to manage the external interface with the environment, while the women managed the internal interface of the home. Neither could encroach on the other's space or domain. Power, authority and decision-making regarding community, caste, and social/political issues rested exclusively with the males.

The social design of this phase evolved a new space for men and women. Slowly but steadily, women's role shrank into a limited life space with rigid boundaries. These boundaries had varied impact on the various strata of women belonging to different socio-economic and caste structures. Women began to acquire a role space which lost its dignity, respect and value.

Many researchers in different parts of the country have carried out authentic studies on the participation of women in family decision-making.

Arya (1964) reported that younger group of male family heads depended more on female family members for decision-making. As age of male family heads advanced there was a greater tendency to consult sons and hence the importance of female (wife) decreased. However, the female members specially the wife, continue to influence decision-making regarding home improvement, irrespective of age of family head. In her study, Puri (1968) revealed that consultation of women in decision-making was the highest in respect of marriage of children followed by children's education and it was the least in the cases of loans, and farm related tasks. According an ILO (1975) study most of the decisions about the allocations of resources within the household are made by men and earnings from increased crop yields are also controlled by them.

Nagre (1976) stated that participation of women in decision-making was the highest (81%) in storage and sale of produce. Women participated jointly with their husbands in decision-making about cropping pattern only in 15 per cent of the cases. Women's involvement was meagre (2.5%) in decision-making on choice of seeds, use of chemical fertilisers, adoption of plant protection measures (use of insecticides and pesticides). Same conclusions were arrived at by Jhansi's (1981) findings that decision-making related to the main occupation, i.e., agriculture was made by male members of the family and women were generally decision-makers regarding the subsidiary enterprises like maintenance of livestock etc. further support the contention. In case of different socio-economic status Yadav (1982) reported that farm women in higher socio-economic status did not enjoy as much involvement in decision-making process regarding farm operations, as those of middle or low socio-economic status families. Regarding the role played by women in decision-making Bajwa (1984) noted that decisions relating to children's education, marriage and dowry were taken jointly by the husband and wife, women had practically no say in decisions relating to saving, loan transactions and purchase of land. Similar findings have also been reported by Ghosh (1987). Kaur (1987) reported that a very high percentage of women played role in the decisions made regarding payment of dowry, food given to expectant and nursing mothers and storage of farm produce. However, they were not involved in decisions taken regarding education of children, taking loans for consumptive expenditures, expenditure pattern of family and farm related activities except storage and sale-purchase of animals.

Farm women's involvement in decision-making regarding farm and home related tasks does not seem to be uniform across the rural women of varying socio-economic strata. Rural women in general, do have certain things in common, still there are intra-economic status variations. The following discussion will throw light on this aspect of Variability in women's role and identity.

Rural Women

Rural women have lived with scarcity of resources and the threat of economic deprivation from last more than 150 years. In many cases, where caste and family taboos are not strong, women joined the secondary stream of the work force in order to raise supplementary resources. They go out of their homes to work in fields, forests and in other settings. Those who can not go out engage in economic activities at home. Women are generally allocated menial jobs and often work for others as paid labourers. Research suggests that, across the centuries, the day-to-day life of rural women has been full of hardship, deprivation and the struggle to survive. Apart from supplementing resources through work, these women have also had to bear the exclusive burden of managing all the activities at home. This monotonous and demanding role and the routine of deprivation is quite typical of rural women belonging to lower down in the caste hierarchy. Women closer to the urban areas seek jobs as construction labourers and household servants or engage in small trades like selling vegetables. For the women, life has meant that they are born to live as beasts of burden and to die without having lived one moment for themselves. Their life is meant to be lived for others, to be sacrificed at the altar of society's discrimination and allocation of a lower status to women. It is a life laid waste both physically and psychologically. Surrendering to the lust of men, their whithering flesh bears child after child, while they live an insecure life under the constant threat of being cast aside at the whim of men. Robbed of dignity, pride and self-worth, women see themselves as prisoners or slaves doomed to obedience and conformity. A woman finds release and freedom only in death.

Upper Class Women

These women live mostly indoors, once they enter the homes of their husbands, they rarely go out. However, within the walls of their home, within the feudal system of a large joint family, run parallel themes of exploitation, intrigue, and counterintrigue, all revolving around the control of resources through legacy and heritage. This is the only life they know as wives. Women from the upper classes are trained for this role right from birth. In such homes, the eldest women of the family holds the authority, the power and the significant social status. The other women of the family, who are younger in age and in the social hierarchy, can only grumble and conform but have no voice of their own. There does not seem to be any significant difference between the role of women belonging to the lower economic groups and those of the upper classes. Women in both categories share the fate of producing children year after year, and encountering sexual exploitation either within or outside the family. The misery of the upper class women is more poignant as she cannot express her anger in public or make a scene. Poorer women at least have the psychological freedom to occasionally vent their frustration in public.

Middle Class Women

Women belonging to the middle classes are squarely caught in the shifting sands of culture. They are neither restricted to the home nor permitted to fulfil their aspirations. They live in a half-way house life *Trishanku*. The legitimacy of their role lies in managing the home, children and husband. Authority over children, a certain amount of economic control on resources, some participation in decision-making, freedom to determine religious observances and finally, the freedom to meet with friends in her peer group, both in enjoyment as well as in a cooperative effort to furnish a yearly supply of storable food items, are the anchors of her well-being. But behind all this, she carries the same taboos as the women of other two classes.

However, one must recognise that the women of this class are perhaps the most significant source of family integration and survival and of the transmission of culture. They are the unsung heroines who, while enduring stress, pain, anguish and frustration, continue to inspire and mobilise their men and children towards the achievement and fulfilment of their dreams. They are the anchors of continuity and ensure the persistence of the ethos–in short, they are the source of cohesiveness of the society.

In this class, the majority of women in traditional agrarian society lived life which was narrow and bounded by prescriptive roles and traditions. Many women from the middle classes and the elite were socially powerful and could perhaps have broken away from tradition, but they rarely took this drastic step.

The industrial and technological ethos began when the roles ascribed to men and women had already crystallised into distinctive sets of activities. In terms of the availability and utilisation of system resources, caste and occupational roles were neatly identified, categorised, wrapped up and sealed. There was little space for negotiation or change. Society was governed by the 'should' and 'musts' of a prescriptive era.

Sex roles distinctions were also evident in terms of occupations, with male more likely to be involved in pursuit requiring strength and technical skills and women in tasks requiring skills in child bearing, home-making and other such operations. However, in case of the rural population of lower and lower middle socio-economic status, women do perform some of the tasks related to farming. Chaudhry and Sharma (1961) revealed in their study that the rural women participate in agricultural operations like manuring, weeding, land preparation, hoeing, sowing, taking care of the standing crops, applying manure and fertilisers, harvesting, threshing, storage, carrying the produce from farm to home and animal care. Similar findings have also been reported by Naine *et al.*, 1966; Singh, 1968; Devadas *et al.*, 1972; Bhatt, 1975; Dhar, 1978; Verma, 1978; Devi and Reddy, 1984.

Another study by Thangamani (1971) reported that women's participation was high in respect of sowing, transplanting, storing and supervision of labour–particularly women labour. Vijayalakshmi (1976) and Sen (1988) were also of the same view. In an exploration made into the women's performance of animal husbandry related tasks. Puri (1975) reported that bringing fodder from the field, chaff cutting, giving water and food to cattle, bathing and cleaning cattle and making cow dung cakes were done by the women only. This was supported by Varma and Malik, 1984; Munjal, 1984; Kaur, 1987 and Dhiman and Khirwar, 1986.

Community wise women's participation in agriculture has been reported to be the highest among tribes, followed by scheduled caste. An analysis made by Deepali (1979) in her study revealed that farm women participated to a large and considerable extent in 11 out of the total 17 farming operations (taking all crops). Out of these eleven, time devoted in five operations ranged from 62 to 66 per cent (gap filling, manure and fertiliser application, harvesting, threshing, winnowing and rodent control).

Anandalakshmy and Kelkar (1980) revealed that during the harvesting season, utilisation of women labour is maximal. Carrying bundles of harvested crop to the spot where threshing would be done is a heavy task which is performed mostly by women. Each bundle is carried as a head load by women, mostly bare footed walking over the sharp shrubs of the harvested fields. Men rarely participated in this task.

In the same way, Kanjla (1980) brought to light that the tasks which were exclusively performed by women were preparing milk products, milking the cattle, cooking food for calves, care of pregnant and milch animals. Dixon (1982) revealed that ploughing and irrigation were man's tasks, sowing may be restricted to men or women or shared, weeding and transplanting are women's tasks, harvesting is frequently shared, and most of the post-harvest operations (threshing, winnowing, drying, husking, cleaning, storing) are performed by women. (Gandhi, *et. al.*, 1986; Rai, 1975). Aggarwal (1983) reported that women provide on an average 70-80 per cent of labour for transplanting, 70-85 per cent for weeding, over 60 per cent for harvesting. They do all the husking and other important roles in seed selection and storage. Bajwa (1983) recorded that women from farming families participated more in storage, sowing, kitchen gardening, care of implements, storage of fertilisers and insecticides, composting and preparing produce for the market than those of non-farming families. Saikia (1983) worked out that work participation rates of females were 58.37 per cent in marginal farms, 57.7 per cent in small farms and 53.91 per cent in medium farms. Agriculture is the most important sector promoting employment to 97.54 per cent of the female workers both in primary and secondary occupations. Out of the total female workers engaged in farm activities, 84.38 per cent worked full time and only 15.62 per cent were engaged as part time workers. Saini (1983) studied that women managed alone the household tasks like work in the kitchen, care of the house, care of children, religious activities etc. The women from farming families spent much more time in the kitchen than those of non-farming families who were mainly the labour class women, who were either paid in cash or kind. In addition, the sowing and harvesting period required much more time of the home maker than the slack period.

Sridevi (1984) revealed that women from agriculturally progressive villages spent on an average 8.2 hours per day on farm as compared to 7.6 hours per day by women farm workers, could work for about 250 days in a year in the progressive village compared to about 210 days in the less progressive village.

Saxena and Bhatnagar (1985) found that participation of women in farm activities was more or less same for tribal as well as non-tribal farm women (about 8.25 hours per day).

Dak *et al.* (1986) revealed that majority of farm women played monopolising or dominant role in 8 out of 17 agricultural tasks–four of them are tending cattle, collecting fodder, and selling livestock products.

Laxmi Devi (1987) reported that the low income category of rural women performed more work in farm activities while the high income category performed more work in home activities. However, the overall time spent was nearly the same for both the classes. Rural women spent on an average 40.4 per cent and 15.8 per cent respectively in home and farm related activities. The remaining 43.8 per cent time was utilised for sleeping, leisure time activities etc. Lepcha (1987) found that average daily hours spent by women in agricultural and livestock production activities were about 7.2 hours per day. Satapathy (1987) found that participation of farm women, especially in respect of small holding, was quite substantial in respect of all important animal related activities. Involvement of women in milking and marketing was independent of land holding systems. Anandalakshmy and Sawhney (1988) concluded that women play a prominent role in crop cultivation, post-harvesting operations, food management and management of milch cattle and milk products. All women in rural areas, irrespective of their age, size of their family, size of land holding, caste and community, perform major agricultural tasks. Women of landless labour families are the poorest, bearing a heavy burden of work in cultivation and performing even the task of land preparation, along with all other tasks.

Hira Nand *et al.* (1988) observed that 62 per cent farm women devoted more than 8 hours per day for farm work during the peak period and 76 per cent of them devoted 4 to 6 hours per day during the slack season in Simla. However, on plains majority of women spent 6 to 8 hours per day and 2 to 4 hours per day during peak and slack seasons (in Bhojpur, Bihar), respectively.

Impact of Industrialisation and Education

With the emergence of the industrial era, which coincided with the independence of the country, a whole new process of work ethics, technology and, consequently, a new life-style was introduced. Apart from fighting against the existing social attitudes backed by 4,000 years of the agrarian ethos, what women essentially needed to do was to confront the concept of the role they were creating for themselves. They had to ask themselves how they were going to design and define their new roles in new era.

Through education, women from the better off socio-economic strata became aware of the existing social disparities and of the condition of women in general, particularly of those belonging to socially and economically deprived groups. It is therefore, imperative that from time-to-time societies, cultures, systems and individuals realign and redefine themselves in ways that are relevant to contemporary needs. Men and women need to respond to the call of their being to generate new meanings, create new landmarks, pave new paths and envisage new patterns of relationships. Agriculture, is however, modernised in our country due to technological advancement. As soon as the agricultural operation is mechanised that work is taken up by men and women labour is thrown out of work or engaged in labour intensive jobs. This has been documented by many studies. As Singh (1972) while discussing about female work participation, green revolution and mechanisation in Punjab held that with adoption of pumpsets for irrigation, wheat threshers, tractors and wheat reapers, female labour was likely to be thrown out of employment. This fact has been envisaged by Apte, 1979; Majumdar, 1975; Boserup, 1977 and Ahuja, 1979. According to ILO (1975) modernisation has not particularly favoured women in most developing countries including India. Gains from agricultural modernisation, technological advancement and dairy employment have benefited male more than women. Sawhney and Devnani (1976) concluded that with mechanisation the hired labour input increased by 38 per cent whereas the family labour input decreased by 50 per cent. Majumdar (1980) revealed that modernisation of agricultural sector has led to mass retrenchment of female agricultural labourers and accentuating of inequalities. It was women who did most of the weeding, harvesting, threshing and winnowing operations before the advent of pesticides, weedicides and various labour saving devices. Pratibha (1981) found that mechanisation did not result in much change in pattern of employment of hired women labour except that work in land preparation and weeding was reduced. Threshing and winnowing was handled by men because of the use of mechanical equipment. Therefore, the level of employment of women went down.

Agriculture is however, getting rapidly modernised in two ways in our country firstly in mechanisation with accelerated installation of tubewells, pumpsets, diesel engines and secondly by the improved agriculture technologies like high yielding variety (HYV) seeds, fertilisers, insecticides and pesticides (Punia, 1980). Mechanisation is used in one or two operations in almost all the crops like maize, wheat, cotton, *Bajra* and rice etc. In crop cultivation, there is mechanisation in different operations by the use of rice hullers, combine harvester; weedicides, pesticides, HYV, tractors, rice transplanter etc. (Aggarwal, 1983; Whyte 1978; Stroller, 1977; Whyte and Whyte, 1978).

Yet, improved agricultural technology accompanied by full mechanisation has negatively affected the women from lower castes, lower income groups who were either landless or had small land holding by reducing the demand for employment in peak season as reported by Kaur (1986). She further added that whenever, employment opportunities for all workers are in short supply, women are most likely to be affected and thus suffer the adverse consequences. Sunita (1986) concluded that the task performed by males are getting mechanised yet the women continue to toil in the labour intensive jobs. The engagement of hired labour was highest in transplanting followed by loading and unloading, harvesting, threshing, winnowing, uprooting and weed control. These operations are not mechanised fully. Thus, mechanisation in paddy cultivation has not made much of dent in hired women labour replacement.

Women Identity in India

The birth of a female child in India evokes responses which vary from feeling let down, disappointed and burdened, anxious about the expenses that will be incurred at her marriage, wondering what her-in-laws would be like and an overall sense of indifferent acceptance; to a sense of joy and contentment and a feeling of being blessed. The parent's indifferent acceptance, of her makes the female child realise that her status is secondary to that of the male child. She experiences no space for herself and learns to be invisible, obedient, confirming and careful about-creating no stress. Thus, the female child experiences herself as being a liability and a burden. She has no clear cut anchors for her psychological location either within herself or in the system. Carving out a space and a meaning for herself without bowing to the well-defined prescriptive roles, is not a risk, she can easily take.

The Indian female child grows into adulthood and wisdom more quickly than the male child. Her experience of attaining physical maturity and the message received from the social world around, tend to converge in the girl's mind and create in her the hope of achieving an ideal relationship in marriage. The female child is aware that hopes of continuing the family and being its economic resource rest with the son. She has already witnessed her mother's life in which creativity was suppressed, potentials denied and her economic dependence emphasised. The social phenomenon of education still confronts women with their secondary and dependent status. Even after they complete their education and embark on a career, women continue to be regarded as being incapable of taking decisions or of exercising authority and as being unstable in jobs.

Contemporary Indian women experience their life-space as a battle ground between the prescriptive roles based on idealised models of a bygone era and the emerging cognitive map of modern society which pulls them towards wider horizons, caught between the traditional past and a future inspired by their own dreams and aspirations, Indian women walk on a tight rope. They carry the burden of both traditional and modern role expectations, yet are denied the privileges of both.

Childhood, schooling, puberty, adolescence and higher education are the landmarks of a female child's life-space. Her life-space during this period is largely occupied by significant family members.

The life space of a female child is directly filled up by the mother and her circumstances. As she grows up, she introjects the core identities from the cultural lore. As she enters, she is exposed to the identities of women in Indian history. She is however, also a witness to living models. Among these models, the ones that stand out clearly are those of women who accept the traditional role of being the shadows of their menfolk yet retain their substantiveness, who are echoes of the prescriptive social system yet retain their voice in times of crisis. They are a source of silent but dependable strength to their families, weathering the storms of family feuds and struggles. They protect their sons and daughters, sometimes their husbands and even whole families from annihilating transaction. In fact, although these women accept the denial of their selves and their social imprisonment with grace, they sow the seeds of new aspirations in their female children.

These women transform their roles into anchors of stability, continuity, security and consistency. In effect, they become the context wherein others derive the autonomy to be themselves. In experiencing the role models of such women, the female child is either aware of or ignores the processes of deprivation and denial of aspiration, meaning and well-being that these women go through. Yet, amidst this, the female child discovers the capacity to regenerate her self from within and derives strength from her commitment to live, to give birth to life and from her capacity to nurture and foster.

For women currently struggling to transcend the inevitable structures of their life-space and create a new space for themselves, this role model of traditional Indian women evokes admiration and awe, but it does not mobilise them to act for themselves because it also evokes pathos and a sense of inadequacy. The struggle to transcend the traditional role is too painful and risky. The universe of the past is familiar, it is easier and safer to navigate in calm and chartered waters.

The emotive map of life role and life space as internalised by the female child from her experience with the mother, father and both their heritages, has many common points with the emotive map introjected by women from the cultural lore. The most common theme is of psychological location. It is always transient. It appears stable and firm but always collapses when needed. A woman's life appears to be like a rudderless boat, drifting, occasionally in serene waters but frequently in turbulent ones. Escape from this drift is possible only by grounding the boat in the swamps of traditional ethos and pathos. Attempts by women to take charge and navigate or to establish anchors, to stop the drift, requires a very strong commitment to deploy their resources on their own and to make adjustments with socially defined situations. Attempts to take charge and navigate beyond the traditional water invokes a wide range of feelings — of anxiety, terror, shame, guilt, joy, hope and dreams. Managing this struggle requires not only conviction but tolerance to withstand many accusations.

Emergence of Role Identity

We now review the three primary locations of women's role identity, i.e.; the adjuster, the asserter, and the drifter, as also some of the variations of these roles.

The Adjuster

Some women having introjected the cultural lore and internalised their experiences with father, mother and their heritage, develop a context for their role. They accept the role of recipient and performer. They accept the construct of duty and role-appropriate feelings and behaviour. They commit themselves to render their best to the family, children and husband with a smiling face. Eventually, these women are seen as being reliable and dependable. By middle age they start acquiring social status and power. They can then influence the choices facing the family and are held in regard. Such women begin to be heard in their system. They also become visible. Essentially, these women become echoes and shadows of ideals promoted by the cultural lore through their transactions with their husband, son, daughter-in-law and grand children.

This heritage is largely of the type which fosters some autonomy and transactional ease in day-to-day living. In this role, these women make themselves available to everybody. But their personal feelings are never stated. They are gracious and earn their dignity. They do not voice any regrets.

The Asserter

Some women, though exposed to the cultural lore, seem to react to their experiences of their mother, father and heritage by internalising assertiveness. They do not like the restrictive life-role and the inevitable nature of their life-space. They try to establish for themselves the role of a navigator and to take charge of their lives. They actively attempt to carve out a space for themselves where they can find expression of their being.

The pathos these women are trying to circumvent or get rid of is still that of being accorded a secondary status, of a lack of psychological location, of being treated as objects and of being restricted to the recipient role. It is interesting to note that most of these women exhibit a tendency to take on the role of patron with a large number of women, specially those with the 'drifter' identity. Very few of these women succeed in retreating and thus creating a personalised world of a wholesome nature.

The Drifter

Besides the introjected cultural lore, the female child was exposed to the constructed lore of a child's life through the writings of popular psychologists and other authors. Thus, both the introjects and the internalisations created a fog in which counterpoints, contradictions, and analogues and identicalities evolved and dissolved from time to time. This process (and its product, the fog) made the problem of psychological and social location all the more difficult and acute. Those with the drifter identity feel deeply disowned and alone. They are restless, desperate to clutch on to something, to experience stability and direction.

One aspect of the universe of women with the drifter identity is anchored in the female child desperately seeking to be affirmed, to belong. The deeper influence of the cultural lore has maintained that this affirmation can be provided by somebody else–God, the father, a guru, the husband and, in modern times, a cause. Caught in the fog of the culture of transcience, the models of affirmation became unstable and diffused.

The adapter, the asserter and the drifter are the three nuclear core identities. They seem, in part, to have continuity with tradition and, in part, are products of the current social and cultural scenario.

Mature women have come to terms with themselves and accept that they are not competing with men but need to create an equation. They anticipate a sharing of responsibilities at home, at work and in living a life of togetherness. There is a need to arrive at a meaningful convergence of their various roles–that of being a traditional *bahu* of being a social asset to the husband, and of carving a career for herself.

Conclusion

Women continue to struggle to get in touch with their substantiveness and to discover for themselves an identity beyond that of a victim and a martyr. In the absence of the system providing any space or role models for such an exploration and being entrenched in the compulsiveness of social roles, they often surrender their search for their personhood and settle down into a reactive mode. Some women have struggled with the system which has often attempted to deploy their potentials both at home and in work setting. They have often confronted within themselves the pathos of the traditional past and the aspired ethos of the present times, but in doing so have overloaded themselves with the processes of both. Most of them have sustained themselves by over-engaging in both roles. Others have fallen either into the role of daughter, or a wife, or a mother. Thus, they have found it difficult to evolve an integrated identity.

Some women continue their struggle to find freedom from dependent, controlling and patronising relationships. Their attempt is to redefine their role to include acting upon the system and to create a legitimate space for themselves. Brought up in the social design of Indian agrarian society, Indian women have over the centuries internalised various processes which they actually detest. They, too, have come to believe that women are like chattels to be owned and protected by a successive set of males; father, brother, husband, son. At no point in time are women encouraged to believe that they are human beings who can be autonomous, can make choices for themselves, and can create life of their own. At best, the only role allowed to them is to adjust, adapt and accommodate to the setting they find themselves in. They are supposed to define or redefine any of these situations through their active participation. They are expected to sacrifice, suffer and, through patience, create a context for the well-being of the family.

Neither the identity of the various princesses nor the role models described above seem to fulfil the basic need that a woman has to find and sustain that are relationship where she can be herself and give free expression to all her potentials without being subjected to aggression, exploitation, criticism, condemnation, and so on. In living out their identities, women have to take on the burden of proving themselves. Yet, in the final analysis, they also have to belong and have to earn social acceptance, and in this process they are once again squarely put back into the role.

Indian women at this juncture get caught up in the dilemma of whether or not to choose the familiar road with its landmarks of *sati* Sita and Savitri. They get caught in the iron shape of the social system which demands a price — that they surrender their existential being so as to acquire a role based security. They have had the courage and the conviction to choose the road of adventure, to discover the vicissitudes of their existence and to grapple with the ardous process of discovering an integrated identity which includes the multiple facets of social and work roles. They have claimed their existential and psychological identity beyond the social roles. Only then have they been able to achieve fulfilment and been able to actualise and experience wider horizons.

In sum, it is believed that women are often victims of their own restricted vision, and Indian women (as also men) still have to discover a third identity which is neither male nor female but human. This identity encompasses the other two identities and is a liberating and life giving force which can revitalise not just women but society at large.

3

RESEARCH METHODOLOGY

This Chapter deals with the research methods and techniques used in the study. The chapter has been organised in six sections:

- I Locale of the study
- II Sampling procedure
- III Selection of the variables
- IV Operationalisation of the variables and their empirical measurements
- V Data collection
- VI Statistical analyses

I. LOCALE OF THE STUDY

The study was conducted in Haryana State. Haryána State has been divided into three zones on the basis of cropping pattern, rainfall, soil type, irrigated area etc. Zone I consists of hot-arid parts of the State (Mohindergarh and Bhiwani districts) while Zone II includes hot and semi-arid region of the State (Sirsa, Hisar, Jind, Gurgaon and Rohtak districts) and Zone III represents irrigated region or hot and humid part (Sonepat, Karnal, Kurukshetra and Ambala district).

Because of hot-arid and lack of irrigation facilities Zone I of the State was excluded for the purpose of study. It was decided to select one district each from other two zones to collect data for this study. Thus, district Hisar from Zone II and district Karnal from Zone III were purposively selected for this study. The main reasons for selection of the two districts were that investigator was well acquainted with the area and area had good communication net work which facilitated approach to the selected villages for collection of data.

For selection of the respondents for this study, one block each from the two districts were randomly selected. They are Hisar block of Hisar district and Karnal block of Karnal district. Again two villages from each of the selected blocks were randomly selected out of the total number of populated villages of the two districts. They are villages of Gogripur, Mehmoodpur of Karnal block and villages of Shahpur and Patan of Hisar community development block. The location of the villages have been depicted in the Map in Fig. 3.1.

The agro-climatic situation of area of study are reported under research setting heading.

Research Setting

In India, the climatic endowment and soil conditions vary considerably between States and regions have tremendous impact on crop-livestock production. A study on involvement of women in different operations under different agro-climatic regions, thus, assumes great importance. Realising this, two districts namely Karnal and Hisar of Haryana State were selected for the present study. The relevant information of the Haryana State and more precisely of the actual study area is described here.

Location

Haryana, one of the northern States of the Country and a part of the Indo-Gangetic plain is situated between 27°03′ to 31°09′ north latitude and 74°05′ to 77°06′ east latitude. Out of the six physiographic regions in the State, upland plain or the Ghaggar Yamuna doab of Karnal and the plain with sand dumes of Hisar district constitute the study area. It is bounded by Uttar Pradesh and Union Territory of Delhi in the east, Rajasthan in the south and south-west and by Punjab in north. The total geographical area of the State is 44.21 lakh hectares. General slope is from the north-east to the north-west and the west. The study was carried out in Karnal and Hisar block of Karnal and Hisar district of Haryana State, respectively.

Climate

Haryana has sub-tropical, semi-dry, continental monsoon's climate. The climate is marked by wide range of temperature and humidity,

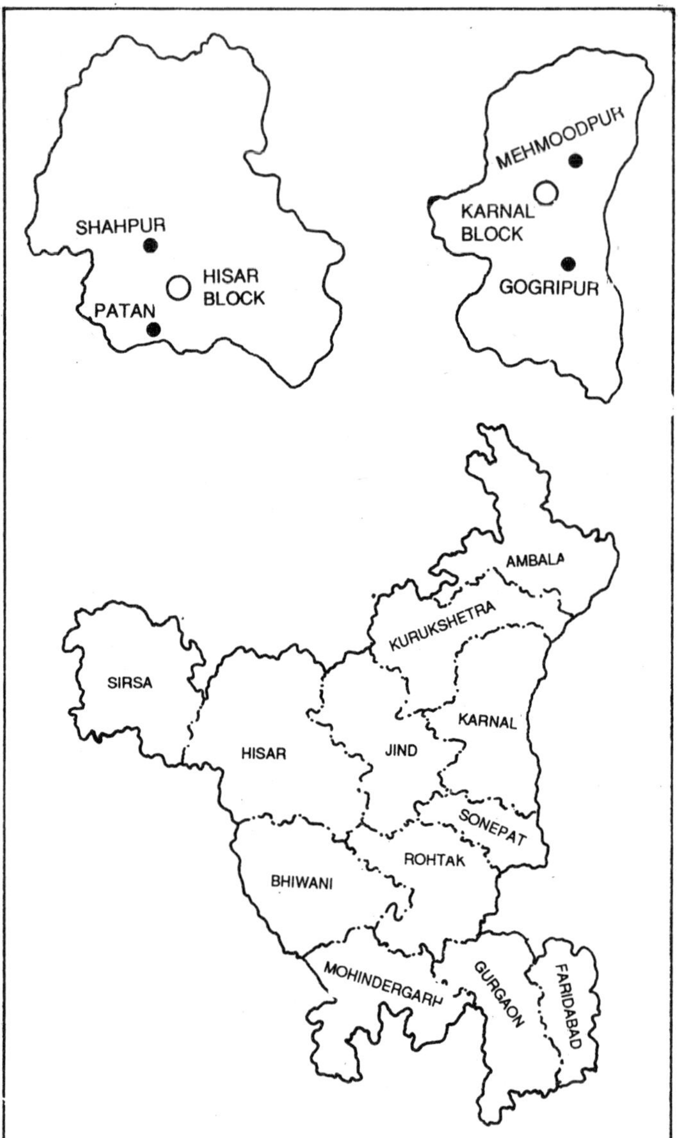

FIG. 3.1: Map of Haryana showing the study area

summers are scorchingly hot but winter are fairly cool. Bio-climatically, the State has three zones and the study area of Karnal falls under hot and humid zone and Hisar falls under hot and semi-arid zone.

The temperature variation in the region as recorded at Ambala, is 3.2°C minimum in January and 47.4°C maximum in June. From February onward, temperatures go on increasing. With the onset of rains, a drop is recorded and when rain ceases the temperature again rises temporarily. But after August, it declines sharply till January and February. There is no lack of sunshine in any part of the year. The normal rainfall in Karnal and Hisar district is 685 mm and 485 mm respectively. The relative humidity is generally low. It varies between 55 in March and 84 in August.

Soil

Soil of the State is characterised as light to medium textured. The loamy-sandy covers 11,755 sq km or 29 per cent of the State area. While sandy-loam is found 22,602 sq km or about 55 per cent of the State area. Soil of Karnal and Hisar is arid-brown and sierozem soil, respectively.

Agriculture

Haryana is predominantly an agricultural and one of the most progressive States in respect of agricultural development. The net available area for cultivation in the State is 4.05 lakh hectare and other uncultivated land including permanent pasture and grassland is 0.47 lakh hectare. Fallow land amount is 1.85 lakh hectare. The total cropped area of the State is 56.88 lakh hectare.

The State has 1.3 lakh hectare area under forest. The area under forest has declined to almost 50 per cent from 1972-75 to 1981-82 (Statistical Abstract of Haryana, 1984-85).

The major source of irrigation in the State are Government canals and tube-wells. The percentage of gross irrigated area to the gross sown area as per 1980-81 records of the State is 60.6 (Statistical Abstract of Haryana, 1984-85).

As per the 1982-83 records, the fertiliser consumption per acre of gross cropped area in Haryana is 47.4 kg as against the national average of 36.6 kg. For the same year the per capita foodgrains production in Haryana is 533 kg as against the national average of only 222 kg. The percentage of area under high yielding varieties of foodgrains in the State is 77. The average crop yield in study area are much higher than national averages. Agriculture is the main while dairy is the subsidiary occupation of the farmers in the State.

Farm operations are carried out either by bullocks or tractors. The use of tractors and thresher is common, even on hired basis. As per the 1980-81 records, there are 76,000 tractors in the State (Statistical Abstract of Haryana, 1984-85).

Agricultural situation in the district of Karnal and Hisar

Karnal: The total cropped area of Karnal district is 5.09 lakh hectare. The percentage of gross irrigated area to the gross sown area as per 1980-81 records of the area is 96.6 (Statistical Abstract of Haryana, 1984-85). The major cereal crops grown in the area are paddy, wheat and maize. Fodder crops are invariably grown twice or thrice in a year. Major fodder crops are maize in *Kharif* and summer and *Berseem* in *Rabi* season.

Hisar : The total cropped area of Hisar district is 8.74 lakh hectare. The percentage of gross irrigated area to gross area sown as per 1980-81 records of the area is 68.9 (Statistical Abstract of Haryana, 1984-85). The major cereal crops grown in the area are wheat, *Bajra*, Mustard and cotton. Fodder crops are grown twice in a year. Major fodder crops are *Jowar, Guar* in *Kharif* and summer and *Berseem* in *Rabi* season.

Animal Husbandry

Haryana is one of the most advanced State of the country in respect of Animal Husbandry. As per the 1986-87 estimates the per day milk production in the State is 5471 thousand litre and per capita milk availability is 474 gm as against the national average 157 gm.

Most of the villages in the study area do not have enough pasture land and hence the practice of leaving animals for grazing is not in vogue. Animal Husbandry in the districts, therefore demand a higher labour investment for cultivation of fodder crop and feeding of the animals.

II. SAMPLING PROCEDURE

The pilot study was conducted in the selected villages to find out the socio-economic status of the villagers with the help of socio-economic status scale developed by Trivedi (1963) with certain modifications which were made in order to suit the local conditions of area of investigation. The socio-economic status scores of all the families of the four selected villages ranged from 4 to 89. The range of scores were divided into three to delimit the three categories of the respondent - high, medium and low. This was done by using the weightage cumulative frequency method. The total weightage cumulative frequency of Gogripur, Mehmoodpur, Shahpur and Patan village were 11,902, 8,095, 5,439 and 7,440 respectively. Total weightage cumulative frequency for each village was divided in three equal parts. The families having cumulative frequency falling in the first part form the group of low socio-economic status. The families having cumulative frequency falling in the second part (Middle part) form the group of medium socio-economic status families. from the remaining families falling under upper part form the group of families having high socio-economic status (Appendix II)

The Sample

As reported in Tables 3.1 and 3.2 the number of families falling in three defined categories in all the four villages were worked out. It was decided to select twenty five families from each of selected villages in proportion to the number of the families in each of the three categories, i.e., low, medium and high. Accordingly, the number of families drawn for collection of data from the three socio-economic status categories from the four selected villages are reported in the last column of Table 3.1 and 3.2. The indicated number of respondents were then randomly selected from the total number of families falling under the particular category in each of the four villages. The total size of sample selected for the study was 100 (25 from each of the four villages).

III. SELECTION OF THE VARIABLES

The relevant variables for the present study were selected on the basis of extensive reviews of relevant literatures, consultations with experts and a pilot study conducted in the area of investigation. The cause-effect relationship between the selected variables were later on hypothesised as per the objectives of the study.

TABLE 3.1: Socio-economic Status of Selected Villages of Karnal District

Sl. No.	Socio-economic status group	Class Interval	Cumulative weighted frequency	No. of families in the category	No. of families included in the sample
		Gogripur Village (Karnal)			
1.	Low	4-25	4-3900	280.00 (54.79)	13.00 (52.00)
2.	Medium	26-36	4550-8090	141.00 (27.60)	7.00 (28.00)
3.	High	37-89	8534-11902	90.00 (17.61)	5.00 (20.00)
	Total			511.00	25.00
		Mehmoodpur Village (Karnal)			
1.	Low	5-22	20-2704	181.00 (52.39)	13.00 (52.00)
2.	Medium	23-33	2957-5381	96.00 (27.42)	7.00 (28.00)
3.	High	34-46	5585-8095	71.00 (20.29)	5.00 (20.00)
	Total			350.00	25.00

TABLE 3.2: Socio-economic Status of Selected Villages of Hisar District

SL. No.	Socio-economic status group	Class Interval	Cumulative weighted frequency	No. of families in the category	No. of families included in the sample
		Shahpur Village (Hisar)			
1.	Low	4-14	12-1910	218.00 (57.67)	14.00 (56.00)
2.	Medium	15-22	2105-3749	103.00 (27.25)	7.00 (28.00)
3.	High	23-59	3933-5439	57.00 (15.08)	4.00 (16.00)
	Total			378.00	25.00

Figures in the parentheses indicate percentage of total respondents.

Contd.

TABLE 3.2 *Contd.*

SL. No.	Socio-economic status group	Class Interval	Cumulative weighted frequency	No. of families in the category	No. of families included in the sample
			Patan Village (Hisar)		
1.	Low	8-26	8-2559	144.00 (52.75)	13.00 (52.00)
2.	Medium	27-39	2856-5210	81.00 (29.67)	8.00 (32.00)
3.	High	40-66	5450-7440	48.00 (17.58)	4.00 (16.00)
	Total			273.00	25.00

Figures in parentheses indicate percentage of total respondents.

Selected Variables and their measurements

	Variables	Measurements
	1	2
1.	Socio-economic status	SES scale of Trivedi (1963) with necessary modifications.
	(a) Caste	SES scale of Trivedi (1963)
	(b) Occupation	SES scale of Trivedi (1963)
	(c) Education of respondent	SES scale of Trivedi (1963) With necessary modifications.
	(d) Social Participation	SES scale of Trivedi (1963)
	(e) Family type	SES scale of Trivedi (1963)
	(f) Family size	SES scale of Trivedi (1963)
	(g) Land holding	SES scale of Trivedi (1963) with necessary modifications.
	(h) Type of House	SES scale of Trivedi (1963) with necessary modification.
	(i) Farm Power	SES scale of Trivedi (1963) with necessary modification.
	(j) Agricultural Implements	Schedule developed for the study.
	(k) Milch animals	Schedule developed for the study.
	(l) Material possession	
	(i) Household appliances	SES scale of Trivedi (1963).
	(ii) Prestige items	Schedule developed for the study.

Contd.

Contd.

	1	2
2.	Age	Chronological age of the respondents in in number of years.
3.	Husband's Education	SES scale of Trivedi (1963) with necessary modifications.
4.	Family Education	SES scale of Trivedi (1963) with necessary modifications.
5.	Total Income	Schedule developed for the purpose.
6.	Expenditure Pattern	Schedule developed for the purpose.
7.	Involvement of Men and Women in selected crop cultivation.	Schedule developed for the purpose.
8.	Involvement of Men and Women in Animal Husbandry.	Schedule developed for the purpose.
9.	Involvement of Men and Women in Household Activities.	Schedule developed for the purpose.
10.	Responsibility	Schedule developed for the purpose.
11.	Accountability	Schedule developed for the purpose.
12.	Knowledge of Modern Crop Technology to Farm Women.	Schedule developed for the purpose.
13.	Drudgery Index	A semi-projective type instrument in pictorial form developed for the purpose.
14.	Impact of Mechanisation on women.	Schedule developed for the purpose.
15.	Contribution of women in cultivation of Major Crops.	Schedule developed for the purpose.
16.	Decision-making Pattern	Schedule developed for the purpose (Puri, 1972).
17.	Prestige Attached with Different Tasks.	Schedule developed for the purpose.
18.	Utilisation pattern of saving.	Schedule developed for the purpose.

IV. OPERATIONALISATION OF THE VARIABLES AND THEIR EMPIRICAL MEASUREMENTS

Here, efforts are made to explicate each general level concept to the operational level to facilitate their measurements.

Socio-economic Status

Chapin (1928) defined that socio-economic status as the position that an individual or a family occupies with reference to the prevailing average standards of cultural possessions, effective income, material possession and participation in the group activities of the community. This definition was operationally used in this study with the assumption that socio-economic status might have had influence on the rural women's involvement in different operations, decision-making, drudgery and impact of mechanisation etc.

Caste

Dhama (1976) defined "caste is a collection of families or a group of families bearing a common name, which usually denotes and is associated with a specific occupation, claiming common descent from a mythical ancestor, human or divine professing to follow the same calling and regarded by those who are competent to give an opinion as forming a single homogeneous community".

Majumdar (1962) has pointed out, that in a country, moored in traditions and dominated by sterotypes, where people are borne to castes and not free to form groups, where individualism has not asserted itself, we cannot ignore institutional matrices and caste structure.

It denotes hierarchial status conferred upon a member of a society from the time immemorial according to law of 'Manu'. Caste was measured by the socio-economic status scale of Trivedi (1963) by assigning weightage:

Schedule caste	1
Lower caste	2
Artisan caste	3
Agricultural caste	4
Prestige caste	5
Dominant caste	6

Occupation

It refers to the main source of livelihood. Occupation was measured by the socio-economic status scale of Trivedi (1963) and scores assigned accordingly.

Education of respondent

Beal and Sibley (1967) have pointed out that the individual's ability to read and write and the amount of formal education, he/she possesses, will affect the manner in which the individual gathers data and relates himself/herself to his/her environment. Operationally education referred to the formal education that a rural housewife had received.

It refers to the academic qualifications of the respondent acquired through formal schooling and training. The years of schooling of respondent was quantified by giving scores as under:

Illiterate	0
Primary	1
Middle	2
High School	3
Technical/Vocational	4
Graduate	5
Post graduate and above	6

Social Participation

Social participation refers to the degree, with which the respondents were involved in formal organisations, as members or office bearers and regularity in their attendance to meetings. Thus, social participation is a voluntary sharing in person to group and group-to-group relationships, beyond the immediate household. It was quantified on the basis of scoring system of the SES scale (Trivedi, 1963).

Family Type

Sahay (1969) stated that family may be nuclear or joint. Nuclear family may be defined as the social group, consisting of married man and woman with their children living together under the same roof and sharing a common hearth. In the present investigation the term nuclear was applied to family units consisting primarily of husband, wife and their children.

Joint family was defined by Sahay (1969) as the social group, consisting of several related individual families, especially those of a man and his sons (in case of patrilineal) or of a woman and her daughters (in case of matrilineal), residing in a single large dwelling. Operationally, the term joint was applied to the families, having at least two married couples, living in common residence, and where the males were related as father, son or as brother-brother and earnings from all sources are pooled together and managed by one family head.

The type of family was measured by the socio-economic status scale (Trivedi; 1963) by assigning scores as under:

Nuclear family	1
Joint family	2

Family Size

Operationally, family size referred to the total number of members in the family consisting of husband, wife, children and other dependents. A family consisting of five members and less was considered small family and more than five members as large family. The score assigned as per SES scale of Trivedi (1963) as:

Up to 5 members	1
Above 5 members acre	2

Land Holding

Land holding refers to the standard acres of cultivated land, possessed by an individual family. The weightage assigned as:

No land	0
1-2.5 acre	1
2.5-5.0 acre	2
5.0-7.5 acre	3
7.5-15.0 acre	4
15.0-25.0 acre	5
Above 25.0	6

Type of House

It refers to the type of house, the respondent possessed at the time of investigation. The weightage assigned as:

Katcha house	1
Mixed house	2
Pucca house	3

Farm Power

It refers to the actual animal draught power possessed by the respondent and used for various farm operations at her farm. The weightage assigned as follows:

1-2 draught animals	1
3-4 draught animals	2
5-6 draught animals and above	3
Tractor	6

Agricultural Implements

In this study, the possession of farm machinery or agricultural implements was measured in terms of quantity of items they possess and weightage assigned to each item. Sum of the assigned weightage of each respondent gave the total score for the agricultural implements. The weightage assigned was as follows:

Up to 4 improved agricultural implements	4
4 and above agricultural implements	5
Tractor driven implements	6

Milch Animals

It refers to the total number of animal heads including dairy possessed by the respondents. The weightage assigned as follows:

1-2 milch animals	1
3-4 milch animals	2
5-8 milch animals	3
9 and above milch animals	4

Material Possession

In this study, material possession refers to the household appliances or items and prestige items possessed by a farm family.

A study of Muse (1954) showed that the addition of household equipment were responsible for the saving of 42 minutes and 762 steps, which was nearly half of the time saved and other half of the steps.

Gross and Gradall (1963) stated that change in hand and body motions, change in work, storage space and equipment and change in the product can improve one's method of doing work. The scoring system was adopted as follows:

	Scores
(i) *Household appliances:* Normal household appliances, viz. cycle, radio, electric fan, coal iron, sewing machine, stove etc. The sum of scores of each item constituted the scores of an respondent for household appliances/items.	1
(ii) *Prestige item:* In the present study, the scores assigned to the prestige items are as:	
Improved household items, viz.; electric iron, electric heater, Mixer blender/grinder, Food processor, juicer, toaster, oven, immerson rod, pressure cooker, improved storage bins, gobar gas plant, solar cooker, sofa set, dining set, double bed, knitting machine, tape recorder/record player, twin player cum recorder, scooter, motor-cycle etc.	2
High prestige items, viz.; Refrigerator, cooler, washing machine, television, video cassette recorder, camera, geyser, car, jeep, matador, telephone etc.	3

The sum of the scores of each item was considered the score of each respondent for prestige item.

Age

It is one of the basic characteristics of an individual linked with his maturity, physical well-being, productivity level and work efficiency. In this study, it refers to the chronological age of the respondent in years at the time of investigation.

Husband's Education

It was measured as education of respondent.

Family Education

It refers to the academic qualifications of all the family members acquired through formal schooling and training. The years of schooling of different family members were quantified by giving scores as per socio-economic status scale of Trivedi (1963). First, each eligible member of the family was given the educational score and then overall educational level of the family was computed by using the formula.

$$\text{Family Education Score (FES)} = \frac{\text{Total education score of the family}}{\text{Total number of eligible members of the family.}}$$

Family Income

In the present investigation family income was referred to as the annual monetary income received by the respondent's family from agriculture, dairy and other sources and was operationally measured in terms of actual family income expressed in rupees.

Expenditure Pattern

Maintenance

It refers to the actual amount spent by the respondent for their livelihood, viz.; food, fuel, cloth, house repair, medicines for family members and cattle. It was measured in terms of rupees. Then it was summed money spent on each item of each respondent.

Development

It refers to the actual amount spent by the respondent on children's education, seeds, fertilisers, irrigation, wage of labourers etc. It was measured in terms of rupees. Then rupees were summed money spent on all items of each individual.

Luxurious items

It refers to the actual amount spent by the respondent on entertainment, smoking (cigarette, Bidi), wine etc. It was measured in terms of rupees.

Then the money was summed up on all the items spent for each individual.

Involvement of Men and Women in Agricultural Activities

In refers to the actual involvement of men and women of different socio-economic strata in terms of actual time spent on different agricultural operations in different selected crops. It was measured with the help of pre-tested structured interview schedule developed for this study.

Involvement of Men and Women in Animal Husbandry Activities

It refers to the actual involvement of men and women of different socio-economic strata in terms of actual time spent on animal husbandry activities. It was measured with the help of pre-tested structured interview schedule developed for this purpose.

Involvement of Men and Women in Household Activities

It refers to the actual time spent by men and women in different household activities. It was measured with the help of pre-tested structured interview schedule developed for this study.

Responsibility

The term responsibility means the State or quality of being responsible for the success or failure of job performed. It was measured with the help of structured close ended questions. The respondents were asked for the responsibility of husband only, wife only, husband and wife both and any other member of family for agricultural operations.

Accountability

This refer to the quality or state of being accountable i.e., bound to give an explanation for or is answerable to non performance on the job. The accountability was measured for each agricultural operation. The scores assigned were:

Very much accountable	5

More accountable	4
Neither much nor less accountable	3
Less accountable	2
Very less accountable	1

Average accountability score is calculated according to his or her responsibility category in different socio-economic strata. This is calculated by the formula:

$$\text{Average Accountability score} = \frac{\text{Sum of each respondents score of accountability in responsibility category}}{\text{Total no. of respondents in each responsibility category.}}$$

Knowledge of Modern Crop Technology to Farm Women

English and English (1958) defined knowledge as a body of understood information possessed by an individual or by a culture. They further explained that knowledge is that part of a person's information which is in accordance with established fact.

Bloom *et al* (1956) considered knowledge as those behaviours and test situations which emphasised that remembering either by recognition or recall of ideas, material or phenomena.

In the present study, knowledge was operationalised as the totality of understood information possessed by a person. It was measured with the help of pre-tested structured interview schedule developed for this purpose. The scores assigned to the responses in dichotomised categories like correct-incorrect. The correct responses were assigned a score of one each and incorrect ones were assigned zero. The total score of each respondent was summed up for each crop separately.

Drudgery Index

In the study, drudgery is operationally defined as physical strains and hardships experienced by human beings. For preparing drudgery index of the farm operations under study a semi-projective type instrument in pictorial form was developed. The advantage of a pictorial form of semi-projective instrument is that it provides a strong stimulus to project one's perception of the drudgery involved in the operation depicted in the picture. It also helps the respondents to make instantaneous and inhibition free response giving vent to his inner reaction. The instrument consisted of ten pictures depicting ten important farm operations, each one of them serve as a stimulus to give expression to the subjective experiences of the respondents with the depicted operations as far as drudgery is concerned (Appendix IV).

The selection of ten operations was made in consultation with a few Agronomists of Indian Agricultural Research Institute, New Delhi and also a few farmers of the area of investigation. The rationale behind the selection was to include all the major operations of cultivation of the most major crops of the area like rice and wheat. The ten operations taken together roughly account for more than 95 per cent of cultural practices involved in cultivation of these crops.

In order to arrive at quantitative measures of the drudgery perceived by the respondents in the ten selected cultural practices, Thrustone's law of comparative judgement was taken resort to. It was postulated that the ten practices depicted in the ten pictures possess varying but unknown degree of drudgery. The ten practices were then ordered on a psychological continuum with respect to the degree of drudgery each one of them involves. For this, all possible pairs of the ten pictures were made which came to be 45 in number $nC^2 = n(n-1)/2$. All the pairs were then presented to each one of the respondents to make a judgement about the degree of drudgery involved in the practices depicted in the pair of pictures in comparative terms, i.e., high-low. Each respondent, therefore, made 45 judgements about which of the two pictures (depicted in a pair of pictures) involve higher degree of physical drudgery. Following the schematic treatment of paired comparison data as suggested by Edwards (1969), the 'F', 'P' and 'Z' matrices were prepared which yielded scores indicating relative degree of drudgery involved in the practices selected for the study (case V Model for complete data in Appendix V).

It may be recalled that there were 30 farm men and 30 farm women respondents who made the comparative judgement of the drudgery involved in the selected practices. The data obtained from these two groups of respondents were analysed separately to yield two continua depicting their judgements of the degree of drudgery involved in the practices. Internal consistency check as suggested by Edwards (1969) was applied to the data obtained form the two set of respondents. This check was evolved to determine whether our observed or empirical proportion (*pij*) agrees with those to be expected in terms of our derived scale values (*pij*¹). The absolute average discrepancy was found to be 0.119 for the male and 0.125 for the female respondents. For a set of ten stimuli the two absolute average discrepancies appear to be on higherside. It was, therefore, decided to apply test of significance to the data as suggested Edwards (1969). For this, the Chi-square test of significance developed by Mosteller (1951b) for discrepancies between observed and expected proportions obtained from case V Method of paired comparison was applied to the two sets of data. This test determine whether the assumptions involved in the case V Model are tenable for a given set of data. The underlying assumptions in case V Model are normality of distribution of discriminal processes, unidimensionality of the psychological continuum and equality of various values of standard deviation of differences as given by Formula 2.3. ($\sigma i - j = \sigma i^2 + \sigma j^2 - 2\ rij\ \ \sigma ij$; Edwards (1969) page 25).

Theoretically, if the test of significance results in rejection of null hypothesis, it might be because of any one or any combination of these assumptions. But in practice, this test is primarily sensitive to lack of unidimensionality. The Chi-square values so worked out, came out to be 167.77 for the male and 199.57 for the female respondents. The reference to Fisher's and Yates table for Chi-square values suggested that the calculated Chi-square values are significant at 0.01 level of probability. Thus the assumptions involved in calculating the scale values under case V Model are not tenable. It was, therefore, decided to resale the values using case III Model as suggested by Edwards (1969).

Before applying case III Model to work out the drudgery index, the stimuli were checked for circular triads. The zeta values calculated for this purpose came out to be 1.00 which indicated non-existence of circular triads in the responses obtained.

Impact of Mechanisation

This refers to a mechanisation as a factor in increasing the productivity of land with greater use of improved technology. In the present study, the degree of farm mechanisation has been operationally defined as the number of labour and energy saving (human and animal) devices possessed by a farmer. It was measured with the help of schedule developed with open and close ended questions as given in Appendix III.

Contribution of Women in Cultivation of Major Crops

It refer to the contribution of farm women per acre of land in cultivation of major crops of farming families only.

Decision-making

Bates (1954) stated "The decision-making process involves a decision-maker (actor), an environment (situation) in which the decision-maker must operate, a set of actions available (means) and a set of goals to be accomplished".

Wilkening (1962) found, that roles of husbands and wives in decision-making were determined more by the perceived needs of farm and household than by culturally determined patterns. These needs varied with the level of economic resources of family which in turn were associated with farm and degree of commercialisation and secondly, with the level of aspirations.

As stated by Reick (1960), decision-making is a mental process, based on conscious reasoning. It is further explained by Rogers (1967), as the process by which an evaluation of the meaning and consequence of alternative line of conduct is made. Craven's study (1963) revealed that farm families of low and medium income were likely to make decisions jointly than families of high income and those who lived on highly specialised farms.

In the present study, it refers to the extent of participation by farm men and women in making decisions in all the important matters of home and farm. it was measured through a schedule developed by Puri (1972).

Prestige Attached to Different Tasks

It refers to the actual performance to task either by wife, husband or jointly. Each task was rated for prestige on a five point continuum as fallows:

More prestigious	5
Less prestigious	4
So-So	3
Less against prestige	2
More against prestige	1

Utilisation Pattern of Saving

It refers to the proper utilisation of the saving. It was measured by structured schedule developed with open ended and close ended questions. The response of respondents were recorded in the categories, viz.; assets building, bank deposits, ornaments, social rites and rituals, children's education, women welfare, narcotics etc.

V. DATA COLLECTION

The whole data were collected in two stages. In the first stage, socio-economic status of all the families of selected villages were measured. The selected respondents were then personally interviewed with the help of a structured schedule. While constructing the main interview schedule, the objectives and dimensions of the study were taken into consideration and statement were included based on the experience, discussion with experts and review of relevant literature. The schedule was pre-tested in a Haryana village (other than the selected villages). The text of the schedules are given in Appendix I, III and IV.

The schedule for farm men and women were filled in by personally interviewed the respondent. The data collection work was done from June to October, 1988. With many respondents, the researcher had more than one sitting.

VI. STATISTICAL ANALYSES

The statistical analytical technique used in this study include as follows:

(1) Frequency
(2) Percentages
(3) Mean
(4) Standard deviation
(5) 't' test
(6) Rank order correlation
(7) Path analysis

4

FINDINGS AND DISCUSSION

The findings of this study are presented under the following heads:

1. Socio-economic profile of women respondents
2. Involvement of men and women in agricultural, animal husbandry and household activities
3. Correlates of work load of men and women for cultivation of different crops
4. Extent of exposure of modern crop technology to farm women
5. Comparative drudgery index of farm men and women for different farm operations
6. Impact of mechanisation on farm women's employment
7. Socio-economic correlates of drudgery
8. Contribution of women in cultivation of major crops
9. Decision-making pattern in the farm family
10. Utilisation pattern of savings.

SOCIO-ECONOMIC PROFILE OF WOMEN RESPONDENTS

As already reported in the chapter on research methodology the data for this study was collected from 100 farm women respondents drawn from high, medium and low socio-economic status. Fifty of them belonged to district Karnal and the remaining fifty to district Hisar of Haryana State. The respondents of the three socio-economic status group from the two districts were compared on the socio-economic variables under study by employing '*t*' statistics. The results are presented in Table 4.1.

TABLE 4.1: Socio-economic Profile of Women Respondents

Sl. No.	Variables	District	Socio-economic Status High (H)	Medium (M)	Low (L)	't' values H-M	H-L	M-L
1	2	3	4	5	6	7	8	9
	Socio-Personal Variables							
1.	Age	Karnal	40.40@ (8.57)	39.79 (5.58)	29.27 (6.93)	0.21	4.04**	4.88**
		Hisar	44.00 (4.21)	37.07 (6.51)	31.74 (8.31)	2.71*	3.99**	2.14*
		Pooled	42.00 (7.20)	38.38 (6.23)	30.53 (7.76)	1.82	5.51**	4.68**
2.	Caste	Karnal	4.40 (0.49)	4.14 (0.99)	2.04 (1.65)	0.75	4.41**	4.35**
		Hisar	4.25 (0.83)	2.47 (1.50)	1.37 (0.62)	3.09**	10.71**	3.35**
		Pooled	4.33 (0.67)	3.28 (1.53)	1.70 (1.28)	2.77**	8.32**	4.97**
3.	Education (Self)	Karnal	0.80 (1.66)	0.36 (0.72)	0.08 (0.38)	0.89	2.12*	1.61
		Hisar	1.50 (1.58)	0.33 (0.70)	0.04 (0.19)	2.47*	4.86**	2.09*
		Pooled	1.11 (1.66)	0.34 (0.71)	0.06 (0.30)	2.19*	4.46**	2.57*
4.	Husband's Education	Karnal	2.50 (1.20)	2.79 (0.86)	0.58 (0.88)	-0.68	5.28**	7.60**
		Hisar	2.88 (1.54)	1.27 (1.69)	0.15 (0.45)	2.34*	8.35**	3.26**
		Pooled	2.67 (1.37)	2.00 (1.55)	0.36 (0.73)	1.49	9.09**	6.51**
5.	Family Education	Karnal	1.33 (0.75)	1.44 (0.56)	0.43 (0.31)	-0.40	5.17**	7.37**
		Hisar	2.13 (0.84)	1.03 (0.70)	0.52 (0.46)	3.35**	7.14**	2.86*
		Pooled	1.69 (0.88)	1.23 (0.67)	0.48 (0.40)	2.02*	7.96**	6.42**
6.	Social Participation	Karnal	0.20 (0.40)	0.07 (0.26)	0.00 (0.00)	0.96	2.61*	1.43
		Hisar	0.75 (0.66)	0.20 (0.40)	0.07 (0.26)	2.50*	4.38**	1.23
		Pooled	0.44 (0.60)	0.14 (0.34)	0.04 (0.19)	2.23*	4.38**	1.70

Contd.

Contd.

1	2	3	4	5	6	7	8	9
7.	Size of family	Karnal	1.70 (0.45)	1.50 (0.50)	1.19 (0.39)	0.99	3.31**	2.14*
		Hisar	1.75 (0.43)	1.47 (0.50)	1.30 (0.46)	1.35	2.49*	1.12
		Pooled	1.72 (0.45)	1.48 (0.50)	1.25 (0.43)	1.66	4.02**	2.25*
8.	Type of Family	Karnal	1.70 (0.45)	1.70 (0.45)	1.12 (0.32)	0.00	4.35**	4.88**
		Hisar	1.75 (0.43)	1.53 (0.50)	1.30 (0.46)	1.03	2.49*	1.56
		Pooled	1.72 (0.45)	1.62 (0.49)	1.21 (0.41)	0.72	4.53**	4.11**
	Economic Variables							
9.	Occupa-tion	Karnal	4.00 (0.00)	3.79 (0.77)	1.88 (1.40)	0.87	4.75**	4.70**
		Hisar	4.00 (0.00)	4.00 (0.00)	2.56 (1.50)	0.00	2.70*	3.71**
		Pooled	4.00 (0.00)	3.90 (0.55)	2.23 (1.49)	0.80	5.03**	5.82**
10.	Type of house	Karnal	3.00 (0.00)	2.43 (0.49)	1.88 (0.75)	2.43*	4.66**	3.63**
		Hisar	2.88 (0.33)	2.53 (0.62)	1.56 (0.83)	1.45	4.35**	3.98**
		Pooled	2.94 (0.23)	2.48 (0.56)	1.72 (0.81)	3.29**	6.32**	4.52**
11.	Land holding	Karnal	3.60 (1.36)	2.20 (0.65)	0.78 (0.89)	3.85**	8.34**	5.73**
		Hisar	5.13 (1.17)	2.27 (0.72)	0.87 (0.98)	6.80**	9.12**	4.44**
		Pooled	4.28 (1.48)	2.24 (0.73)	0.83 (0.99)	6.30**	11.20**	6.76**
12.	Farm power	Karnal	1.60 (0.80)	0.89 (0.56)	0.81 (0.68)	2.94**	2.98**	0.40
		Hisar	1.75 (0.66)	0.91 (0.52)	0.81 (0.39)	3.01**	5.05**	0.91
		Pooled	1.67 (0.75)	0.90 (0.55)	0.81 (0.55)	4.08**	5.18**	0.67
13.	Agricul-tural Imp-lements	Karnal	4.20 (1.66)	1.00 (1.96)	0.15 (0.77)	4.19**	10.07**	1.95
		Hisar	5.50 (0.87)	2.60 (2.06)	1.63 (1.97)	3.78**	5.37**	1.51
		Pooled	4.78 (1.51)	1.83 (2.17)	0.91 (1.67)	5.05**	8.68**	2.14*

Contd.

1	2	3	4	5	6	7	8	9
14.	Milch-Animals	Karnal	1.70 (0.78)	0.93 (0.26)	0.92 (0.38)	3.47**	4.02**	0.05
		Hisar	1.75 (0.66)	0.87 (0.50)	0.67 (0.47)	3.61**	5.20**	1.29
		Pooled	1.72 (0.73)	0.90 (0.40)	0.79 (0.45)	5.00**	6.40**	1.04
15.	House-hold app-liances	Karnal	14.30 (0.00)	12.04 (2.73)	5.68 (4.45)	2.61*	5.93**	4.77**
		Hisar	15.13 (0.76)	9.68 (5.56)	5.50 (6.08)	2.73*	4.42**	2.20*
		Pooled	14.67 (0.65)	10.82 (4.58)	5.59 (5.38)	3.53**	7.10**	4.42**
16.	Pestige-items	Karnal	182.60 (274.43)	8.25 (11.32)	0.29 (1.16)	2.40*	3.47**	3.59**
		Hisar	206.50 (286.00)	18.90 (34.51)	0.00 (0.00)	2.56*	3.89**	2.87**
		Pooled	193.22 (279.88)	13.76 (26.58)	0.14 (0.83)	3.45**	5.09**	3.74**
17.	Total Income	Karnal	611.00 (419.21)	227.50 (70.81)	96.12 (79.01)	3.38**	6.12**	5.19**
		Hisar	1910.00 (855.69)	436.48 (147.26)	191.19 (156.64)	5.48**	10.22**	2.60*
		Pooled	1188.33 (916.35)	335.59 (341.75)	144.55 (133.49)	4.55**	8.15**	3.61**
18.	**Expenditure Pattern**							
(a)	Main-tenance Expen-diture	Karnal	253.58 (122.15)	138.27 (402.50)	74.69 (45.91)	3.61**	6.48**	4.35**
		Hisar	426.75 (246.71)	169.70 (71.83)	112.89 (61.05)	3.81**	6.19**	2.71**
		Pooled	330.54 (206.75)	154.48 (60.79)	94.15 (57.42)	4.32**	7.59**	4.45**
(b)	Deve-lopment Expen-diture	Karnal	240.90 (314.35)	51.34 (29.74)	20.87 (39.02)	2.26*	3.58**	2.55*
		Hisar	368.50 (160.94)	93.63 (66.86)	50.80 (76.04)	5.82**	7.87**	1.82
		Pooled	297.61 (265.39)	73.22 (56.44)	36.11 (62.58)	4.42**	6.73**	2.65**
(c)	Luxury or wasteful Expen-diture	Karnal	8.83 (10.05)	6.03 (4.94)	3.66 (2.39)	0.91	2.50*	2.05*
		Hisar	19.25 (21.63)	5.89 (6.28)	7.71 (9.60)	2.26*	2.19*	-0.66
		Pooled	13.46 (17.05)	5.96 (5.67)	5.72 (7.34)	2.19*	2.68**	0.15

@ Indicate Mean Scores.
Figures in parentheses indicate standard deviation.
* Significant at 0.05 level of probability
** Significant at 0.01 level of probability.

Age

Age of the respondents vary from 20 to 55 years with an average of 35 years. Age of the respondents have nothing to do with socio-economic status. Yet, it is interesting to note that women from high and medium socio-economic status were significantly older than the women of low socio-economic status. The standard deviations reported under parentheses clearly indicate that in case of the respondents of all three socio-economic status, there was a lot of heterogeneity so far as age is concerned. Yet the average, somehow, came to be in the order of socio-economic status levels of the respondents.

Caste

In rural areas, caste is an important social institution. The respondents belonged to different caste which have varying social status. Based on the status, castes were assigned scores. The caste scores obtained by the respondents of the high, medium and low socio-economic status groups (pooled) were statistically compared as reported in Table 4.1. A perusal of the data presented in Table 4.1 points out that there was significant difference between the respondents of high, medium and low socio-economic strata in both the districts, except in case of the caste hierarchy of high and medium socio-economic strata of the respondents of Karnal district. It is thus obvious that the farmer's socio-economic status and their caste hierarchy move in the same direction. The caste composition of the respondents in order of their hierarchy are presented in Table 4.2.

Education (Self)

Education is one of the important factors which accelerates growth and development in farming. There were significant differences between the level of education of the women of high, medium and low socio-economic strata in case of the sample from Hisar district and also the pooled data. However, in Karnal district, the significant difference between educational levels of the respondents was found only between the respondents of high and low socio-economic strata. However, majority of women (55%) of high socio-economic stratum had obtained formal education up to middle class whereas a great majority of the women from the low socio-economic stratum (77%) were found to be illiterate in both the districts. About 50 per cent of the medium socio-economic status women had obtained formal schooling up to middle and matriculation standard in Karnal district. However, in Hisar district, a thumping majority (73%) of women from medium socio-economic status were found to be illiterate.

TABLE 4.2: Distribution Respondents According to their Caste

Sl. No.	Caste	Caste Scores	Socio-economic status								
			Karnal District %			Hisar District %			Pooled data %		
			High	Medium	Low	High	Medium	Low	High	Medium	Low
1.	Jat	5	40.00	35.71	19.23	50.00	20.00	0.00	45.00	27.85	9.62
2.	Brahmin	4	60.00	57.15	7.69	12.50	6.67	0.00	36.25	31.90	3.85
3.	Roar, Arora	3	0.00	0.00	0.00	37.50	6.67	7.41	18.75	3.33	3.70
4.	Kumar (Artisan caste)	2	0.00	0.00	3.85	0.00	33.33	22.22	0.00	16.67	13.03
5.	Scheduled Caste	1	0.00	7.14	69.23	0.00	33.33	70.37	0.00	20.25	69.80
	Total		100.00	100.00	100.00	100.00	100.00	100.00	100.00	100.00	100.00

Husband's Education

The educational level of the respondents is important as discussed earlier but the educational level of their husbands have perhaps greater significance for development since they are the family heads and major decision-makers. From the data reported in Table 4.1, it is clear that educational level of husbands coming from high and medium socio-economic status were significantly higher than those coming from lower socio-economic stratum.

A perusal of the basic data about husband's education revealed that out of the 50 per cent illiterate husbands (total 100) 40 per cent belonged to lower socio-economic status group alone and only 2 per cent belonged to high socio-economic status group. On the other hand out of the 20 per cent husbands who got their education up to high school and above, only one per cent belonged to low socio-economic status group and 10 per cent to the high socio-economic status group followed by 9 per cent to medium socio-economic status group. It is, therefore, obvious that heads of the families of high socio-economic group had higher level of schooling may be because they could afford the expenditure involved.

Family Education

Since farming by and large is a family profession, all members of the family perform various farm related tasks and so are capable of affecting improvement in farming. Educational level of all the members of the family, therefore, is important for acquisition, comprehension and acceptance of information about improved farming. The data reported in Table 4.1 clearly indicate that the respondent's family education follows the trend of other two variables i.e., self and husband's education. When the data were reset to find out the number of family members having some education, the result as reported in Table 4.3.

TABLE 4.3: Frequency Distribution of Respondents According to their Family Education

Socio Economic Status Categories	Frequency of families having educated members in various proportions					
	100% family members	75% to 99% family members	50% to 74% family members	25% to 49% family members	Less than 25% family members	Total
High	3.00 (16.67)	1.00 (5.55)	10.00 (55.56)	4.00 (22.22)	0.00 (0.00)	18.00 (100.00)
Medium	1.00 (3.45)	7.00 (24.14)	10.00 (34.48)	9.00 (31.03)	2.00 (6.90)	29.00 (100.00)
Low	0.00 (0.00)	1.00 (1.89)	9.00 (16.98)	32.00 (60.38)	11.00 (20.75)	53.00 (100.00)
Total	4.00	9.00	29.00	45.00	13.00	100.00

Figures in parentheses indicate percentages.

From the data reported above some interesting results came to limelight. There were only four of the hundred families where all the members received some education, three of such families belonged to high socio-economic status, one to medium and none to low. In case of the next level of education, i.e., 75 per cent and more family members having received some education, there were nine families in high and medium socio-economic status and only one belonged to the low category which constitutes 53 per cent of the total sample. On the other hand, eleven of the thirteen families with less than one fourth of family members having received some education belonged to low socio-economic status category and none to high socio-economic status category. This further confirms that lower the extent of family education, lower is the socio-economic status. The data further confirm the findings that education favours the socio-economic status or *vice versa* in farming community of Haryana.

Social Participation

A perusal of Table 4.1 reveals that social participation of the women respondents of high socio-economic stratum was significantly higher than those of the respondents of medium and low socio-economic strata. There was no significant difference between the social participation of women respondents of medium and low socio-economic strata. The average social participation score obtained by the respondents of all categories are less than one out of the maximum possible score of six. This clearly points to the fact that farm women generally do not take part in social activities and keep themselves away from social institutions. The only institution which has secured some participation of women is *Mahila Mandal*. At least in Hisar district, about one fourth of the respondents of high socio-economic status and 13 per cent of the medium socio-economic status were found to participate in village *Mahila Mandals*. The participation of the women of low socio-economic status however was found to be almost negligible. However, in Karnal district about one-tenth of the respondents of high socio-economic status were found to participate in village *Mahila Mandals*. The participation of the women of medium and low socio-economic strata were found to be almost negligible.

Size and Type of Family

As farming demands extensive labour force, size of family is considered significant in farming community. This is one of the reasons that small family norm has not been able to take root in rural areas. The data reported in Table 4.1 reveal that respondents of higher socio-economic status had greater size of family than the respondents of lower socio-economic status which is contrary to general belief. Further a look into the data reveal that it is so because of existence of number of joint families in the higher socio-economic group. The data in this regard as reported in Table 4.4 clearly reveal that out of 18 respondents belonging to high socio-economic status, 13 had joint families and only 5 had nuclear families. In contrast to this, out of 53 families belonging to low socio-economic status 12 had joint and 41 had nuclear families. Respondents belonging to medium socio-economic status fall in between these two extremes. It clearly indicates that with the declining socio-economic status there is greater tendency of disintegration of joint family system. If we take into account the size of nuclear family which includes husband, wife and their children, the data seem to project the picture in reverse order as reported in Table 4.4.

TABLE 4.4: Frequency Distribution of Respondents According to Type and Size of Family

Category	Socio-economic status							
	High		Medium		Low		Overall	
	Frequency	Average children per family	Frequency	Average children per family	Frequency	Average children per family	Frequency	Average children per family
Nuclear	5	2.60	13	2.31	41	2.37	59	2.37
Joint	13	5.76	16	5.31	12	5.83	41	5.61

Occupation

The data regarding occupation of the respondents have been presented in Table 4.1. A perusal of the data reveals that there was no significant difference between the occupation of high and medium socio-economic status groups of both the districts. However, statistical difference among the occupation of high-low and medium-low socio-economic status groups were observed in both the districts. It is obvious that the respondents of high and medium socio-economic strata were having cultivation as their main occupation in both the districts. Whereas majority of the respondents (60%) of low socio-economic stratum were agricultural labourers others (40%) were found engaged in cultivation of land on share basis in both the districts. Heavy dependence of the respondents on agriculture for seeking out their livelihood was revealed by the data.

Type of house

Type of house is a component determining respondents' socio-economic status. It is natural that three socio-economic status group differ significantly in respect of type of their houses. It is however, interesting to note that overwhelming majority (17 out of 18) of the respondents of high socio-economic status group had *pucca* houses but more than 52 per cent of those belonging to medium socio-economic status group and 21 per cent of those belonging to low socio-economic status group also had *pucca* houses. Even the mixed type of houses were owned by good number of those belonging to low (28%) and medium (45%) socio-economic status group. However, an overall percentage of *katcha* house (27 out of 28) were owned by the respondents of low socio-economic status group only as evident from the data reported in Table 4.5.

TABLE 4.5: Frequency Distribution of Respondents According to their Type of House

	Socio-economic status			
Type of house	(High (%)	Medium (%)	Low (%)	Total (%)
Katcha	0.00 (0.00)	1.00 (3.45)	27.00 (50.94)	28.00 (28.00)
Mixed	1.00 (5.55)	13.00 (44.83)	15.00 (28.30)	29.00 (29.00)
Pucca	17.00 (94.45)	15.00 (51.72)	11.00 (20.76)	43.00 (43.00)
Total	18.00 (100.00)	29.00 (100.00)	53.00 (100.00)	100.00 (100.00)

Land-holding

Since size of land holding is an important component of socio-economic status, it is natural that the respondents of three socio-economic strata groups should significantly differ from each other in respect of this variable. The average size of land holding of the respondents of the three socio-economic status groups were found to be 40.30, 5.43 and 1.40 acres, respectively, which is far apart from each other. It is however, of interest to note that about 22 per cent belonging to high socio-economic status group were small farmers. In case of the respondents of medium socio-economic status group about 21 per cent had marginal land holdings and 45 per cent had small land holdings. However, all the landless respondents included in the study belonged to low socio-economic status group only (Table 4.6).

TABLE 4.6: Frequency Distribution of Respondents According to their Land Holdings

	Socio-economic status			
Categories	High	Medium	Low	Total
Landless	0.00 (0.00)	0.00 (0.00)	28.00 (52.00)	28.00 (28.00)
Marginal	0.00 (0.00)	6.00 (20.68)	12.00 (22.64)	18.00 (18.00)
Small	4.00 (22.22)	13.00 (44.83)	13.00 (24.53)	30.00 (30.00)
Medium & Big	14.00 (17.78)	10.00 (34.48)	0.00 (0.00)	24.00 (24.00)
Total	18.00 (100.00)	29.00 (100.00)	53.00 (100.00)	100.00 (100.00)

Figures in parentheses indicate percentages.

Draft Animal

The number of draft animals possessed by farmers is generally governed by the size of their land holding but it is interesting to note that one farmer from high socio-economic status had no draft animal and 14 of the 28 landless respondents had some draft animal. A further probe in the matter suggested that the high socio-economic status farmers who did not have draft animal is a government employee, hence he was not able to maintain animal and used to get his land cultivated by hiring of tractor from others in the village. The reason why 14 landless respondents had kept either one or two draft animals is that they had kept bullock cart for hiring out to others for transport of farm produce and other material from one place to another. This is a major source of earning for them. It is of further interest to note that fifteen of the respondents owned a tractor and all of them except one also owned some draft animal. The tractors are generally owned by big landholders but even they had kept one pair of bullock for some of the operations which they think is better done by desi plough than by tractor, for example, sowing of fine grained seeds like mustard etc. Two tractor owners had only medium size of land holding (5.0-7.5 acre). An enquiry in the matter suggested that they had made 'hiring out of tractor' a side business to supplement their earnings from farming.

Agricultural Implements

The data regarding agricultural implements are reported in Table 4.1 bring to light some interesting results. Possession of agricultural implements is supposed to go along the size of agricultural implements possessed by the respondents of medium socio-economic status were not found significantly higher than those possessed by the respondents of low socio-economic status. This is largely because the respondents of both the category possessed only some traditional agricultural implements and improved agricultural implements which were assigned higher scores were possessed by only the respondents of high socio-economic status. Out of the fifteen tractor owners thirteen belonged to the high socio-economic status and most of these tractor owners possessed several tractor driven farm implements and other pov· implements like threshers etc.

Milch Animals

The data regarding possession of milch animals by the respondents as reported in Table 4.1 reveal that respondents of high socio-economic status possessed significantly higher number of milch animals than the rest of the respondents. On an average, respondents from high socio-economic status possessed about 3 animals, those from medium socio-economic status possessed 1 milch animal and the respondents from low socio-economic status possessed less than 1, i.e., 0.80 animals. Cultivation of fodder is major constraint in rearing of milch animals. Since the respondents of high socio-economic status have larger size of land holding, they could afford to put larger area under fodder, hence they were able to maintain 2 to 3 milch animals.

In case of the respondents of medium socio-economic status, their size of land holding is not big enough to afford sufficient land for fodder cultivation, since they have to grow crops for their own consumption which is the first preference. This is possibly the reason for their low possession of milch animal. On this logic, possession of milch animals by the respondents of low socio-economic status should have been much less than what they are since a sizeable proportion of them do not hold any land. But in case of respondents of this category, selling of milk is an important source of earning and hence they maintain some milch animals even with a strain for arrangement of fodder.

Household Appliances

Possession of household appliances is a direct function of economic standing of the household. This is the reason that three categories differ significantly from each other in possession of household appliances. The respondents from high socio-economic status were found to possess household appliances of worth Rs. 1,467 on an average, whereas those from medium socio-economic status possessed household appliances worth Rs. 1,082 and the respondents of low socio-economic status worth Rs. 559 only. It may however be mentioned that possession of household appliances by low socio-economic status is bare minimum like essential cooking utensils without which they cannot manage their survival. some of the household appliances which are required for better and efficient living like cycle and radio was mostly possessed by the respondents of high socio-economic status to some extent by those of the medium socio-economic status.

Prestige Item

Items included under this are those household appliances and possessions which adds prestige to the family in the society. Although most of these items increase work efficiency and makes the living comfortable, they are treated as prestige items in the culture because of the grinding poverty which has engulfed a large part of the society in which they manage themselves without these items and with no regret. This includes items like scooter/motorcycle, refrigerator, television, electric iron, electric heater, mixer blender/grinder, food processor, juicer, toaster, oven, pressure cooker, improved storage bins, gas *chullah,* sofa set, double bed, knitting machine, cooler and things like that. The data reported in Table 4.1 clearly reveal that prestige items were possessed mostly by the respondents of high socio-economic status. Their average possession came to worth over Rs. 19,000 as compared to the average possession of respondents of medium socio-economic stratum which was only worth over Rs. 1300 and that of the low socio-economic stratum group which was worth barely Rs. 14. It is obvious from the data that the concept of efficient living has not reached in the families who have medium level of economic standing.

Total Income

The data regarding total income as reported in Table 4.1 clearly indicate that there was vast disparity in income among the respondents of various socio-economic strata. The average family income of the respondents drawn from high socio-economic stratum though came out to be Rs. 1,19,000 per annum, it was only about Rs. 33.5 thousands for the respondents of medium socio-economic stratum and 14.5 thousand for respondents of low socio-economic stratum. Even though family size varies, this variation fail to substantially reduce the gap in income. Even the average per head income in case of respondents from high socio-economic stratum was found to be over Rs. 11.5 thousand per annum. It was only about Rs. 3.8 thousand for the respondents of medium socio-economic stratum and about Rs. 2.3 thousand from low socio-economic stratum. There was none from the high and medium socio-economic strata respondents who fell below poverty line but 16 of 53 respondents of low socio-economic stratum was found to fall below poverty line. It may however be noted that there were 28 respondents from low socio-economic stratum who were landless and at least 12 of them were found to be above poverty line. Reason for this was that some of them was permanent labourers at Haryana Agricultural University farm and others had some supplementary source of income like dairying, bullock cart for hiring out and so on. Some other farmers of this category who had smaller holding of land used to lease in others land which substantially increased their earning.

Expenditure Pattern

The total expenditure made by the respondents were broadly divided into three categories. A part of the expenditure is necessary for self-maintenance which includes expenditure on food, cloth, fuel and medicine. Another category of expenditure includes those made for developmental purpose like children's education, improved farming and cattle rearing. The third category of expenditure includes those made on things like smoking, liquor, etc. which are avoidable and hence termed as wasteful expenditure. In addition, there are some expenditure on miscellaneous items which were not taken into account for the purpose of present discussion. The data reported in Table 4.1 reveal that the expenditure made by the respondents of three socio-economic status group significantly differ in most of the cases. However, the data do not give detailed information on expenditure pattern. The data of expenditure, therefore, with details of the item of expenditure are reported in Table 4.7. Which reveal that in absolute term of amount spent by the respondents from high socio-economic stratum was the highest followed by the medium and low socio-economic stratum group respondents.

TABLE 4.7: **Expenditure Pattern of Respondents**

Sl. No.	Expenditure on different items	Socio-economic status		
		High	Medium	Low
1.	Maintenance Exp.			
(a)	Food	16844.44 (14.17)	7737.93 (23.06)	4926.41 (34.77)
(b)	Cloth	6633.33 (5.58)	2689.31 (8.01)	1681.13 (11.70)
(c)	Fuel	1500.00 (1.26)	400.00 (5.17)	300.00 (6.09)
(d)	Medicine	2323.33 (1.95)	1686.20 (5.03)	975.47 (6.75)
2.	Development Exp.			
(a)	Education	2777.77 (2.34)	692.07 (2.06)	192.26 (1.33)
(b)	Farming per acre of land	1973.77 (1.66)	1293.44 (3.86)	923.91 (6.73)
(c)	Dairying	7253.53 (6.10)	3286.86 (9.79)	1641.50 (11.36)
3.	Wasteful Expenditure	1346.00 (1.13)	596.00 (1.77)	572.00 (3.96)

Figures in parentheses indicate percentages of income.

The data reported in parentheses are quite revealing. Whereas about 14 per cent of income is spent on food by the respondents of high socio-economic status group, it is as high as about 35 per cent for the respondents of low socio-economic status and this trend seems to continue for the other three items of this category, i.e., cloth, fuel and medicine. It is, therefore, clear that the respondents from lower socio-economic status had to spent a good amount of their earning on food, cloth, fuel and medicine in order to maintain themselves. With the increase in socio-economic status there is a corresponding increase in expenditure on these items which adds to the quality of living, but that increase is not in proportion of the increase in income. A similar trend is also discernible in case of development expenditure made on farming and dairying. But the trend is just the reverse in case of the expenditure made on children's education. Not only the lowest amount but lowest proportion of income is spent on children's education in case of low socio-economic status group and both the amount and proportion increases with the increase in socio-economic status. This clearly reveals that respondents from higher socio-economic status value education more and are more prone to spent a higher amount of their earnings on this. Therefore, it is important to inculcate the value of education in those coming from low socio-economic status in order to set desirable pace of development in the society.

A more serious situation appears to exist in case of wasteful expenditures. The amount on an average spent on such wasteful habits like smoking and drinking etc. by the respondents of low socio-economic status is very close to the amount spent by those from medium socio-economic status but substantial higher in percentage of the earning made by them (3.96 against 1.77). In case of high socio-economic status respondents, this percentage is still lower (1.13). This clearly points to strong need of some educational programme to make the people of low socio-economic status aware of this wasteful expenditure and its ill-effects on their social, economic and health status. Saving from this expenditure can be gainfully deployed in some development expenditure like children's education and things can dramatically change within a short span of time.

INVOLVEMENT OF MEN AND WOMEN IN AGRICULTURAL, ANIMAL HUSBANDRY AND HOUSEHOLD ACTIVITIES

The data reported in Table 4.8 indicate that in case of wheat, cotton and *Bajra* cultivation, there are six operations namely pesticide dusting, spade work during field irrigation, ploughing, sowing, manure and fertiliser application and marketing of grains, which were performed exclusively by men. There is only one operation namely keeping part of grains for consumption which was performed exclusively by women. The remaining six operations namely weeding by *kasola*, weeding by *khurpi*, carrying load on head, winnowing, harvesting and threshing are such which are performed by both men and women. Out of these six operations, carrying head load is predominantly performed by women (over 80%). The other operations viz.; weeding by *kasola* and *khurpi* (over 80%) in wheat, weeding by *kasola* in *Bajra* (81.30%) and in cotton (51.11%) crops is predominantly performed by women. Even in case of harvesting about 65 per cent work load in wheat and *Bajra* crops and more than 90 per cent in case of cotton crop is taken up by women. The heavy pre-occupation of women in these six operations have resulted in substantially higher labour contribution by them in cultivation of wheat, cotton and *Bajra* crops. On an average a woman devotes 14.72, 28.42 and 15.37 man days against 14.19, 19.83 and 12.85 man days devoted by a man in cultivation of one acre of wheat, cotton and *Bajra* crops, respectively. The data reveal that major pre-occupation of women labour in wheat, cotton and *Bajra* cultivation is in weeding and harvesting. They spend more than half of their man days (deployed in wheat, cotton and Bajra cultivation) only in these operations which they carry either with the help of *khurpi, kasola* or sickle. These appear to be exclusive domain of women closely followed by 'on head carrying' of the harvested crop. In case of paddy cultivation same operations are dominated by men exclusively with addition of uprooting of seedlings whereas transplanting and keeping part of grains for consumption were found to be the exclusive domain of women. The remaining four operations, i.e., weeding by *khurpi*, carrying head load, harvesting and threshing were performed by both of them but man days devoted by women heavily outnumbered the man days devoted by men. Weeding and carrying harvested crop on head were predominantly (to the extent of 90%) performed by women. Even in case of harvesting and threshing, about 70 per cent of the operations are performed by women. The heavy pre-occupation in these six operations of paddy cultivation had shown higher contribution of labour by women than by men. On an average farm women devote 23.02 man days as against 19.18 man days devoted by men in

TABLE 4.8: Farm Operation Undertaken by Men and Women in Wheat, Paddy, Cotton and Bajra Cultivation in terms of Man Days

Sl. No.	Operations	Wheat (N=100)		Paddy (N=50)		Cotton (N=50)		Bajra (N=50)	
		Men	Women	Men	Women	Men	Women	Men	Women
1	2	3	4	5	6	7	8	9	10
1.	*Male Dominated Operations*								
(a)	Pesticide dusting	36.81 (100.00)	-	20.74 (100.00)	-	88.39 (100.00)	-	10.00 (100.00)	-
(b)	Spade work during field irrigation	199.30 (100.00)	-	349.62 (100.00)	-	94.65 (100.00)	-	50.48 (100.00)	-
(c)	Field Preparation	316.75 (100.00)	-	80.76 (100.00)	-	166.25 (100.00)	-	114.00 (100.00)	-
(d)	Nursery raising/ sowing	131.25 (95.99)	5.5 (4.02)	25.00 (100.00)	-	66.75 (100.00)	-	58.50 (100.00)	-
(e)	Manure and lizer application	135.25 (100.00)	-	65.88 (100.00)	-	49.57 (100.00)	-	30.01 (100.00)	-
(f)	Uprooting of seedling		-	100.00 (100.00)	-	-	-	-	-
(g)	Marketing of grains	93.88 (100.00)		32.00 (100.00)	-	44.75 (100.00)	-	33.51 (100.00)	-
2.	*Female Dominated Operations*								
(a)	Weeding by '*Kasola*'	57.00 (13.67)	360.00 (86.33)	-	-	376.00 (48.89)	393.00 (51.11)	46.00 (18.70)	200.00 (81.30)
(b)	Weeding by '*Khurpi*'	35.00 (11.22)	277.00 (88.78)	21.00 (9.59)	198.00 (90.41)	-	-	-	-
(c)	Carrying load on head	22.50 (13.98)	138.50 (86.02)	8.00 (8.08)	91.00 (91.92)	10.00 (13.33)	65.00 (86.67)	20.00 (20.00)	80.00 (80.00)
(d)	Transplanting	-	-	-	300.00 (100.00)	-	-	-	-
(e)	Winnowing	-	-	-	-	-	-	39.00 (31.20)	86.00 (68.80)
(f)	Keeping part of grains for consumption	-	72.00 (100.00)	-	19.25 (100.00)	-	-	-	27.25 (100.00)
3.	*Jointly Operated Operations*								
(a)	Harvesting	280.00 (35.04)	519.00 (64.96)	126.00 (31.27)	277.00 (68.73)	28.00 (4.24)	632.00 (95.76)	171.00 35.04)	317.00 (64.96)
(b)	Threshing	247.63 (71.23)	100.00 (28.77)	124.00 (31.79)	266.00 (68.21)	67.00 (16.83)	331.00 (83.17)	69.88 (54.64)	58.00 (45.36)
	Total	1418.62	1472.05	953.00	1151.25	991.36	1421.00	642.38	768.45
	Average	14.19	14.72	19.18	23.02	19.83	28.42	12.85	15.37

Figures in parentheses indicate the percentage of the total man days deployed in the operation by men and women.

cultivation of one acre of paddy. A similar finding has also been reported by Aggarwal (1983) that in rice systems women supply on an average 70-80 per cent of labour for transplanting, 70-85 per cent for weeding, over 60 per cent for harvesting. They do all the husking and have important roles in seed selection and storage. Domination of female in paddy transplantation is also reported in the studies conducted by Devadas *et al.* (1972), Chakravarty (1975), Chatterjee (1975), Saikia and Gogoi (1981), Panikar (1983), Gulati (1984) and Kaur (1987).

Similarly, harvesting, weeding and winnowing were also the major jobs of womenfolk engaged in agriculture. Studies by Chaudhry and Sharma (1961), Singh (1968), Devadas *et al.* (1972), Mulay (1974), Dhar (1978), Pandey *et al.* (1988) and Kaur (1987), conducted in different parts of India also support the present finding.

In a study Devi and Reddy (1984) observed that activities like ploughing, sowing, fertilser application, plant protection measures and irrigation were male jobs.

Attempts were made in this study to find out the significant difference in the involvement of men and women of various socio-economic strata in selected crop cultivation. The mean scores of the involvement (in terms of man days) of rural men and women in wheat cultivation with '*t*' values have been incorporated in Table 4.9. The results reveal that in male dominated operations (ploughing, sowing, manure and fertiliser application, spade work during field irrigation, pesticide dusting and marketing of grains) the mean scores of man's work load in all the three socio-economic strata were higher than women. However, in case of female dominated operations (weeding by *kasola*, weeding by *khurpi*, carrying load on head, keeping part of grains for consumption) and jointly operated operations (Harvesting and threshing) the mean scores of women's involvement were quite higher than that of men in high, medium and low socio-economic strata as well as in pooled data. The data further unfolds significant differences between men and women's work load in male dominated, female dominated and jointly performed operations. The overall work load of women (in terms of average man days) was higher than that of men in high and medium socio-economic strata whereas, reverse trend was observed in case of low socio-economic stratum. However, pooled data show higher involvement of women that men in wheat cultivation. The statistical results show no difference between men and women's involvement in wheat cultivation in all the socio-economic strata categories.

TABLE 4.9: Involvement in Various Operations - Inter Sex Variation (Wheat)

Sl. No.	Operations	Socio-Economic Status	Man days		't' value
			Male (Mean Score)	Female (Mean Score)	
1.	Male dominated operations	High	8.88	0.00	14.58**
		Medium	8.90	0.06	22.71**
		Low	9.44	0.06	32.21**
		Total	9.18	0.05	41.13**
2.	Female dominated operations	High	0.56	7.63	-5.37**
		Medium	0.38	8.32	-7.84**
		Low	1.29	8.85	-13.09**
		Total	1.14	8.48	-15.29**
3.	Jointly operated operations	High	3.88	6.82	-5.88**
		Medium	3.92	6.00	-7.15**
		Low	4.38	6.17	-6.67**
		Total	5.27	6.20	-5.19**
4.	Overall operations	High	13.31	14.44	-0.71
		Medium	13.20	14.49	-1.13
		Low	15.12	14.86	0.35
		Total	14.24	14.68	-0.74

** Significant at 0.01 level of probability.

The Table further shows that the average work load of men and women of low socio-economic stratum were higher than those in case of high and medium socio-economic strata respondents in all the operations. It may, therefore, be inferred that low socio-economic stratum women performed more work in crop cultivation than the women of high and medium socio-economic strata. The obtained involvement pattern confirms heavy work load of women of low socio-economic stratum, probably due to their being landless and wage earners. As these womenfolk are left with no alternative for their livelihood they are pushed into less skilled, low paying, more time consuming and drudgery prone activities. Similar findings have also been reported by Pandey *et al.*, (1988) and Kaur (1987).

Table 4.10 indicate that, in case of paddy cultivation, the overall mean score of the involvement of farm women was more than men in high, medium and low socio-economic strata as well as in pooled data. There was significant difference between involvement pattern of men and women in paddy cultivation. It may, therefore, be inferred that women from different socio-economic strata spend more time (in term of man days) than men in paddy cultivation. Amongst other operations, the male dominated operations (ploughing, puddling, manure and fertiliser application, nursery raising, spade work during field irrigation, pesticide dusting, uprooting of seedlings and marketing of grains) the mean score of men was substantially higher than women in all socio-economic strata categories. However, in case of the female dominated operations (transplanting, weeding by *khurpi*, carrying load on head, keeping part of grains for consumption) and jointly operated operations (harvesting and threshing) the mean score of women's involvement was much higher than men in high, medium and low socio-economic strata as well as pooled respondents. The significant difference was obtained between men and women's involvement in male dominated, female dominated and jointly operated operations. The data further confirm that the average work load of low socio-economic stratum women was higher than high and medium socio-economic strata women in all the different operations. It may be inferred that low socio-economic stratum women performed more work in cultivation of paddy crop than high and medium socio-economic strata women. This may be mainly due to economic compulsions of sustaining their families for which they have to overburden themselves by agricultural work and that too in low paying, less skilled and monotonous activities. Similar findings have also been reported by Pandey *et al.* (1988) and Kaur (1987).

TABLE 4.10: Involvement in Various Operations - Inter Sex Variation (Paddy)

Sl.No Operations	Socio-Economic Status	Man days		't' value
		Male (Mean Score)	Female (Mean Score)	
1. Male dominated operations	High	14.06	0.00	14.41**
	Medium	11.89	0.00	17.89**
	Low	14.16	0.00	25.42**
	Total	13.50	0.00	32.10**
2. Female dominated operations	High	0.00	9.85	-13.20**
	Medium	0.36	12.07	-11.55**
	Low	0.88	13.11	-14.34**
	Total	0.56	12.16	-20.12**
3. Jointly operated operations	High	3.80	12.40	-5.82**
	Medium	5.64	10.07	-4.96**
	Low	5.15	10.69	-9.39**
	Total	5.02	10.86	-11.34**
4. Overall operations	High	17.86	22.25	-2.08*
	Medium	17.89	22.14	-2.94**
	Low	20.20	23.80	-2.70*
	Total	19.08	23.02	-4.25**

* Significant at 0.05 level of probability.
** Significant at 0.01 level of probability.

Involvement of men and women in various operations of cotton cultivation are reported in Table 4.11 which reveal that the overall involvement of female (in terms of average man days) was quite high than male in high, medium and low socio-economic strata as well as in case of total respondents. There was significant difference between men and women's work load in cotton cultivation. It may, therefore, be inferred that women's involvement in cotton cultivation was higher than men. In male dominated operations women's involvement was negligible in different socio-economic strata as well as among pooled sample.

TABLE 4.11: Involvement in Various Operations - Inter Sex Variation (Cotton)

Sl.No Operations	Socio-Economic Status	Man days		't' value
		Male (Mean Score)	Female (Mean Score)	
1. Male dominated operations	High	9.50	0.00	19.81**
	Medium	8.93	0.00	23.23**
	Low	11.13	0.00	22.34**
	Total	10.21	0.00	30.49**
2. Female dominated operations	High	6.50	8.25	-1.66
	Medium	7.73	8.53	-0.75
	Low	8.07	9.77	-2.24*
	Total	7.72	9.16	-2.60*
3. Jointly operated operations	High	0.00	17.62	-23.40**
	Medium	4.13	18.07	-5.52**
	Low	4.22	20.41	-15.65**
	Total	2.90	19.26	-16.79**
4. Overall operations	High	16.00	25.87	-6.78**
	Medium	20.77	26.60	-2.30*
	Low	20.83	30.18	-6.57**
	Total	19.83	28.42	-7.36**

* Significant at 0.05 level of probability.
** Significant at 0.01 level of probability.

The mean score of women's involvement was higher than men in high, medium, low and pooled data in female dominated operations (weeding by *kasola*, carrying load on head). In jointly operated operations (harvesting and threshing), the mean score of women's work load was more substantial than men in all socio-economic strata categories and in pooled data. This Table 4.11 further unfolds that the average work load of low socio-economic stratum women was higher than those of the high and medium socio-economic strata and among pooled data. It may, therefore, be concluded that low socio-economic stratum women performed extensive work in crop cultivation than high and medium socio-economic strata women. This may be due to economic compulsions of earning their livelihood through working in field as agricultural labourer. They seldom hesitate to work even on low wages. Similar results were reported by Pandey *et. al.*, (1988) and Kaur (1987).

Involvement of men and women in various operations of *Bajra* cultivation is summarised in Table 4.12. The mean score of women's work load was higher than men's work load in high, medium and low socio-economic strata as well as in the pooled data. There was significant inter-sex variation in high-low socio-economic strata and among the pooled data. As evident from the data, in case of male dominated operations, women's involvement was nil. Whereas, in case of female dominated operations (weeding by *kasola*, carrying load on head, winnowing and keeping part of the grains for consumption) males' involvement was substantially and significantly lower in case of all the socio-economic strata categories of respondents in isolation as well as in combination. Even in case of jointly operated operations, there was significant inter-sex variation among high, medium and low socio-economic strata and pooled data. Obviously, therefore, women undertake more burden of work in *Bajra* cultivation than men. Table 4.12 further shows that the low socio-economic stratum men and women do more work as compared to high and medium socio-economic strata respondents. This may be due to the fact that the respondents of low socio-economic stratum are mainly landless and wage earners. They are left with no alternative for their livelihood, therefore, perform more work on other's field as agricultural labourers, that too in low paying, less skilled and monotonous work in crop cultivation. Similar results were reported by Kaur (1987).

Per day participation of men and women (in terms of average man days) in various activities of animal husbandry is reported in Table 4.13 with inter-sex variations ('*t*' values). The data reveal that in case of high socio-economic stratum men scored over women in all animal care related activities viz.; bringing fodder from field, chaffing the fodder, feeding the animals, protect animals from ticks, cleaning of animals, milking and offering water. It ranged from 0.03 to 0.65 man days in case of men as against only 0.01 to 0.27 man days in case of women. The inter-sex differences in these activities were found to be statistically significant at 0.05 level of probability except in case of preparation of feed, cleaning and milking.

TABLE 4.12: Involvement in Various Operations - Inter Sex Variation (Bajra)

Sl. No.	Operations	Socio-Economic Status	Man days: Male (Mean Score)	Man days: Female (Mean Score)	't' value
1.	Male dominated operations	High	5.35	0.00	11.22**
		Medium	5.03	0.00	20.59**
		Low	6.59	0.00	8.68**
		Total	5.93	0.00	13.59**
2.	Female dominated operations	High	1.37	7.78	-15.91**
		Medium	3.33	7.97	-8.65**
		Low	3.37	7.83	-6.79**
		Total	2.28	7.86	-19.20**
3.	Jointly operated operations	High	3.31	8.12	-8.03**
		Medium	4.31	7.57	-6.96**
		Low	6.42	6.67	-4.03**
		Total	4.34	7.17	-9.33**
4.	Overall operations	High	14.00	15.90	-2.20*
		Medium	15.53	15.60	-0.11
		Low	10.67	14.50	-6.78**
		Total	12.85	15.37	-5.79**

* Significant at 0.05 level of probability.
** Significant at 0.01 level of probability.

In case of both medium and low socio-economic strata, women's involvement in animal husbandry activities were found to be significantly higher than men in all the operations except cleaning of animals in case of medium and offering water in case of low socio-economic strata respondents.

The results, therefore, clearly reveal that animal husbandry is predominantly a male affair in case of the farmers of high socio-economic status, whereas, it is predominantly a female affair in case of the farmers of medium and low socio-economic status. However, the researcher's experience suggests that this inter socio-economic status variation is quite deceptive. If we take only family labour into account, the result would be uniform across the socio-economic status. But in case of the farmers of high socio-economic status, a majority of them had employed permanent male labour to look after the animals which has in fact off set the results in favour of men in this stratum. Similar results were reported by Puri (1975) that bringing fodder, chaff cutting, preparing and giving feed to cattle, bathing, cleaning the cattle Shed, milking, making curd, butter and *ghee* were wife centered activities. This was supported by Haque (1968), Singh (1968), Devadas *et al.*, (1972), Chaudhry (1973), Ahuja (1979), Sinha (1980), Awasthy (1982), Varma and Malik (1984), Munjal (1984), Dhiman and Khirwar (1986), Kaur (1987).

TABLE 4.13: Involvement in Various Operations - Inter Sex Variation (Animal Husbandry)

Sl.No	Operations	Socio-Economic Status	Man days: Male (Mean Score)	Man days: Female (Mean Score)	't' value
1.	Bringing Fodder from field	High	0.25	0.08	3.32**
		Medium	0.06	0.17	-3.98**
		Low	0.00	0.20	-16.61**
		Total	0.06	0.17	-6.27**
2.	Chaffing the fodder	High	0.10	0.02	3.93**
		Medium	0.03	0.06	-2.11*
		Low	0.00	0.08	-16.26**
		Total	0.03	0.06	-4.83**
3.	Prepare feed for animals	High	0.07	0.05	0.93
		Medium	0.02	0.05	-3.64**
		Low	0.00	0.05	-15.31**
		Total	0.02	0.05	-5.93*
4.	Protect animals from ticks	High	0.03	0.01	2.74*
		Medium	0.00	0.01	-3.91
		Low	0.00	0.01	-4.24
		Total	0.01	0.01	0.0[illegible]
5.	Cleaning of animals	High	0.07	0.05	1.0
		Medium	0.04	0.05	-1.21
		Low	0.01	0.05	-6.30**
		Total	0.03	0.05	-3.10**
6.	Milking	High	0.06	0.04	1.22
		Medium	0.00	0.05	-10.96**
		Low	0.00	0.04	-9.69**
		Total	0.01	0.04	-5.95**
7.	Offering Water to animals	High	0.08	0.03	2.00*
		Medium	0.02	0.07	-3.25**
		Low	0.03	0.05	-1.16
		Total	0.04	0.05	-1.41
8.	Overall Operations	High	0.65	0.27	3.01**
		Medium	0.17	0.46	-6.49**
		Low	0.06	0.48	-15.05**
		Total	0.20	0.44	-6.17**

* Significant at 0.05 level of probability.
** Significant at 0.01 level of probability.

Involvement of women of different socio-economic strata in household activities is presented in Table 4.14. The mean score of women's work load in slack season was higher than their work load in peak season in high, medium and low socio-economic strata respondents as well as in case of the total sample. There was significant difference between the work load of women in peak and slack seasons. It may, therefore, be inferred that the household work is predominantly performed by women irrespective of their socio-economic status and women devote more time in household activities during slack season. On an average a woman devotes about 6 hours in peak season against 7 hours in slack season. It means women devote more time in slack season which is mainly due to the reason that women have more time at their disposal to look after the children and house properly, which they tend to neglect in peak agricultural season. Similar findings have also been reported by Sinha (1980), Jately (1981) Khan and Ayesha (1982), Manohar (1983) and Kaur (1987).

TABLE 4.14: Involvement of Women in Household Activities

Sl.No	Operations	Socio-Economic Status	Man days		't' value
			Peak season mean score	Slack season mean score	
1.	Household activities	High	0.89	0.98	-2.13*
		Medium	0.79	0.90	-9.09**
		Low	0.66	0.85	-6.09**
		Total	0.74	0.89	-5.88**

* Significant at 0.05 level of probability.
** Significant at 0.01 level of probability.

Responsibility and Accountability of Male and Female for Various Farm Operations

The data regarding responsibility and accountability are reported in Table 4.15. The responsibility has been worked out in percentage and accountability has been worked out as described in the chapter on research methodology. The responsibility and accountability as reported in Table 4.15, reveal that there are male dominated operations

TABLE 4.15: Responsibility and Accountability of Male and Female for Various Farm Operations

Sl. No.	Operations		SOCIO-ECONOMIC STATUS					
			Responsibility			Accountability		
			High	Medium	Low	High	Medium	Low
1	2		3	4	5	6	7	8
Male Dominated Operations								
1.	Field Preparation	Joint	00.00(00.00)	00.00(00.00)	00.00(00.00)	00.00	00.00	00.00
		Husband	18.00(100.00)	29.00(100.00)	53.00(100.00)	3.06	2.63	2.66
		Wife	00.00(00.00)	00.00(00.00)	00.00(00.00)	00.00	00.00	00.00
2.	Sowing	Joint	00.00(00.00)	2.00(6.90)	3.00(5.66)	00.00	3.00	3.00
		Husband	18.00(100.00)	27.00(93.10)	50.00(94.34)	3.11	2.59	2.69
		Wife	00.00(00.00)	00.00(00.00)	00.00(00.00)	00.00	00.00	00.00
3.	Manure and fertilizer application	Joint	00.00(00.00)	00.00(00.00)	00.00(00.00)	00.00	00.00	00.00
		Husband	18.00(100.00)	29.00(100.00)	53.00(100.00)	3.06	2.77	2.55
		Wife	00.00(00.00)	00.00(00.00)	00.00(00.00)	00.00	00.00	00.00
4.	Plant Protection measures	Joint	00.00(00.00)	00.00(00.00)	00.00(00.00)	00.00	00.00	00.00
		Husband	18.00(100.00)	29.00(100.00)	53.00(100.00)	2.89	3.14	3.00
		Wife	00.00(00.00)	00.00(00.00)	00.00(00.00)	00.00	00.00	00.00
5.	Irrigation	Joint	00.00(00.00)	00.00(00.00)	00.00(00.00)	00.00	00.00	00.00
		Husband	18.00(100.00)	29.00(100.00)	53.00(100.00)	3.11	3.27	3.08
		Wife	00.00(00.00)	00.00(00.00)	00.00(00.00)	00.00	00.00	00.00
6.	Marketing of grains	Joint	00.00(00.00)	00.00(00.00)	00.00(00.00)	00.00	00.00	00.00
		Husband	18.00(100.00)	29.00(100.00)	53.00(100.00)	3.28	3.14	2.65
		Wife	00.00(00.00)	00.00(00.00)	00.00(00.00)	00.00	00.00	00.00
Female Dominated Operations								
7.	Weeding	Joint	01.00(5.56)	7.00(24.14)	11.00(20.75)	2.83	2.78	2.64
		Husband	6.00(33.33)	0.00(0.00)	2.00(3.78)	3.36	0.00	3.40
		Wife	11.00(61.11)	22.00(75.86)	40.00(75.47)	3.00	3.25	3.40

Contd.

TABLE 4.15 *Contd.*

1	2		3	4	5	6	7	8
8.	Transplanting	Joint	01.00(05.56)	5.00(17.24)	9.00(16.99)	2.00	2.62	3.36
		Husband	5.00(27.78)	4.00(13.79)	8.00(15.09)	3.28	3.25	3.55
		Wife	12.00(66.67)	20.00(68.97)	36.00(67.92)	4.00	3.00	3.00
9.	Thinning	Joint	2.00(11.11)	7.00(24.14)	10.00(18.87)	2.75	2.83	2.47
		Husband	5.00(27.77)	4.00(13.79)	7.00(13.21)	3.44	3.14	2.40
		Wife	11.00(61.11)	18.00(62.07)	36.00(67.92)	3.00	2.80	2.90
10.	Gap filling	Joint	2.00(11.11)	7.00(24.14)	10.00(18.87)	2.83	2.83	2.59
		Husband	5.00(27.77)	4.00(13.79)	7.00(13.21)	3.62	3.28	2.82
		Wife	11.00(61.11)	18.00(62.07)	36.00(67.92)	3.00	2.80	2.69
11.	Sun drying	Joint	00.00(00.00)	00.00(00.00)	00.00(00.00)	00.00	00.00	00.00
		Husband	00.00(00.00)	00.00(00.00)	00.00(00.00)	00.00	00.00	00.00
		Wife	18.00(100.00)	29.00(100.00)	53.00(100.00)	3.07	3.08	3.15
12.	Keeping part of grains for consumption	Joint	00.00(00.00)	00.00(00.00)	00.00(00.00)	00.00	00.00	00.00
		Husband	00.00(00.00)	00.00(00.00)	00.00(00.00)	00.00	00.00	00.00
		Wife	18.00(100.00)	29.00(100.00)	53.00(100.00)	3.60	3.62	3.57
13.	Taking bundles to threshing floor	Joint	5.00(27.78)	10.00(34.48)	12.00(22.64)	2.80	2.92	2.88
		Husband	2.00(11.11)	00.00(00.00)	00.00(00.00)	2.62	0.00	0.00
		Wife	11.00(61.11)	19.00(65.51)	41.00(77.36)	3.08	3.06	3.08
Jointly Operated Operations								
1.	Harvesting	Joint	12.00(66.67)	29.00(100.00)	53.00(100.00)	3.19	3.14	3.05
		Husband	6.00(33.33)	00.00(00.00)	00.00(00.00)	3.47	0.00	0.00
		Wife	00.00(00.00)	00.00(00.00)	00.00(00.00)	00.00	00.00	00.00
2.	Threshing	Joint	12.00(66.67)	29.00(100.00)	53.00(100.00)	3.00	2.94	2.89
		Husband	6.00(33.33)	00.00(00.00)	00.00(00.00)	3.47	0.00	00.00
		Wife	00.00(00.00)	00.00(00.00)	00.00(00.00)	00.00	00.00	00.00

Figures in parentheses indicate percentages.

viz.; field preparation, sowing, manure and fertilizer application, plant protection, irrigation and marketing of grains, in all socio-economic strata. The accountability in field preparation for high socio-economic stratum was 3.06 which was followed by 2.63 and 2.66, respectively of the respondents of medium and low socio-economic strata. The accountability of the respondents in male dominated operations ranged from 2.89 to 3.28 in high socio-economic stratum, 2.59 to 3.27 in medium socio-economic stratum and 2.55 to 3.08 in case of low socio-economic stratum. The female dominated operations viz.; weeding, transplanting, thinning, gap filling, sun drying, keeping part of grains for consumption and taking the bundles of harvested crops to threshing floor were the responsibility of wife (over 60%) in all socio-economic strata. The accountability of female respondents in female dominated operations ranged from 2.00 to 4.00 in high socio-economic status, 2.62 to 3.62 in medium socio-economic status and 2.40 to 3.52 in case of low socio-economic stratum. The jointly performed operations of harvesting and threshing were the joint responsibility of husband and wife in all socio-economic strata. The accountability of joint respondents ranged from 2.89 to 3.47 in all socio-economic strata. It can however, be concluded that the accountability is more in case of jointly operated operations followed by female dominated operations and male dominated operations.

CORRELATES OF WORK LOAD OF MEN FOR CULTIVATION OF MAJOR CROPS — WHEAT AND PADDY

Table 4.16 presents correlation and path coefficients of the nine independent variables with work load of men in cultivation of wheat and paddy crops. None of the '*r*' values was found to be significant. However, age had very substantial positive direct effect (0.2153). The results therefore, indicate that aged farmer has a tendency to undertake heavier work load than the younger ones. The possible reason seems to be that aged farmers were both physically and mentally settled in farming, whereas, many of the younger farmers even if settled in farming are not that much involved in farming and they continue to have several other interests due to which they often visit cities or other places outside the village which obviously reduced their work load in farming. However, there are two variables which channelled substantial indirect effects due to which, the correlation coefficient of this variable was reduced to the level of non-significance. These variables are self-education and type of family. Their direct effects were also found to be quite substantial and negative. This indicates that educated and those coming from joint families have a tendency to undertake less work load than those who are less educated and coming from nuclear families. In case of joint family, working hands are more and hence the total work load is shared among them making it relatively lighter for an individual. The reason why educated farmers had shouldered relatively lower farm work load are as follows:

(a) Some of the educated farmers were in government and semi-government services and hence they were able to work on the farm only beyond their official duty hours.
(b) A number of educated farmers involve themselves in social activities and social institutions like *Panchayat*, co-operatives and youth club etc. in which they devote some time and to that extent they were free from farming which reduced their work load.
(c) All or atleast most of the non farm activities of a family like making various purchases from the nearby town/market place, maintaining social relations within and outside the village etc. are undertaken by those family members who have received some education. Therefore, the working hours of educated farmers in the family on the farm has reduced their involvement in non-farming activity which reduces their work load.

Correlates of Work Load of Women for Cultivation of Major Crops — Wheat and Paddy

The data related to the correlates of work load undertaken by women in cultivation of two major crops wheat and paddy are reported in Table 4.17. As the results reveal none of the correlates of women's work load was found to yield significant correlation coefficient but the data of direct effect, channelled by some of the correlates to the work load, make some interesting revelations. They are as follows:

Age

The correlation coefficient of the variable age is very small in magnitude but negative which indicates a tendency on the younger farm women to undertake heavier farm work load. The direct effect of

TABLE 4.16: Correlates of Work Load of Men for Cultivation of Major Crops - Wheat and Paddy

Sl. No.	Independent variables	'r' values	Direct effect	Total indirect effect	Substantial Indirect effect through
1.	Age (X1)	0.1944	0.2153	-0.0209	-0.0870(X3), -0.0606(X6), 0.0541(X7).
2.	Caste (X2)	0.1392	0.0095	0.1297	0.1213(X4), -0.1018(X3), 0.0691(X7), -0.0536(X6).
3.	Self Education (X3)	-0.0061	-0.1707	0.1645	0.1098(X1), -0.0761(X6), 0.0536(X4).
4.	Family Education (X4)	0.0280	0.0744	-0.0464	-0.1229(X3), -0.0716(X6), 0.0670(X1).
5.	Social Participation (X5)	-0.0127	-0.0016	-0.0111	-Nil-
6.	Type of Family (X6)	-0.0203	-0.1332	0.1129	0.0979(X1), -0.0975(X3).
7.	Land holding (X7)	0.2039	0.0879	0.1159	0.1325(X1), -0.0815(X3).
8.	Farm Power (X8)	0.1911	0.0809	0.1102	0.0825(X1), 0.0534(X7).
9.	Total Income (X9)	0.1400	0.0362	0.1038	0.0784(X1), -0.0593(X3), -0.0592(X6), 0.0525(X8), 0.0515(X7).

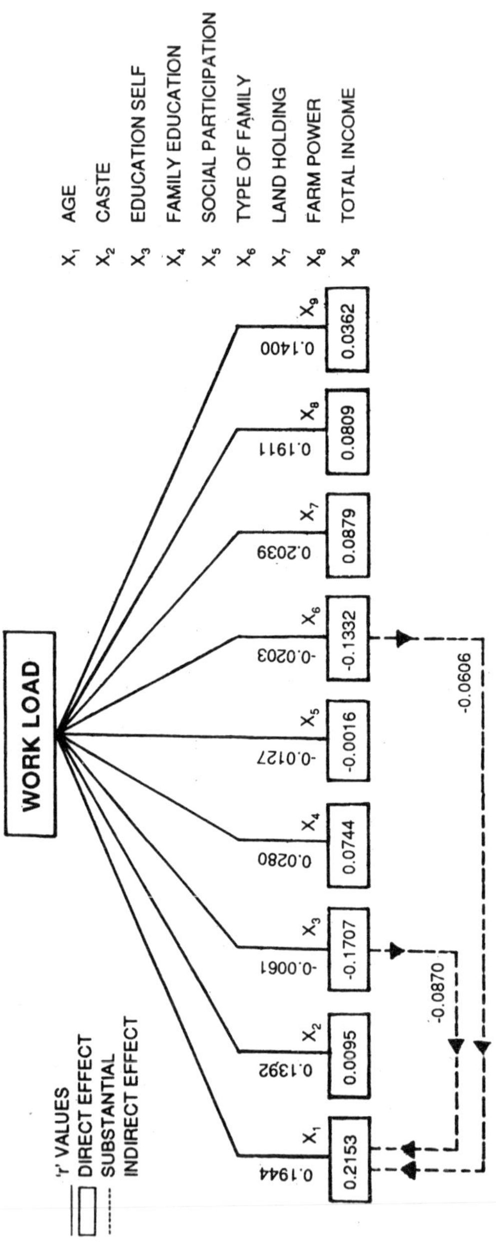

FIG. 4.1: Correlates of work load of men for cultivation of major crops-wheat and paddy

TABLE 4.17: Correlates of Work Load of Women for Cultivation of Major Crops - Wheat and Paddy

Sl. No.	Independent variables	'r' values	Direct effect	Total indirect effect	Substantial Indirect effect through
1.	Age (X1)	-0.0083	0.1809	-0.1892	-0.1595(X8), -0.0955(X4), 0.0656(X6).
2.	Caste (X2)	-0.0509	0.0766	-0.1275	-0.2178(X8), -0.1594(X4), 0.1073(X1), 0.0971(X6).
3.	Self Education (X3)	0.1490	0.4413	-0.2923	-0.1708(X8), -0.1630(X4).
4.	Family Education (X4)	-0.1599	-0.3265	0.1666	0.2203(X3), -0.2119(X8), 0.0793(X6), 0.0529(X1).
5.	Type of Family (X5)	-0.1146	-0.0550	-0.0596	-0.1975(X8), -0.1695(X4), 0.1173(X3), 0.0807(X1), 0.0595(X5).
6.	Land holding (X6)	-0.0995	0.1342	-0.2337	-0.3528(X8), -0.1929(X4), 0.1581(X3), 0.0885(X1), 0.0554(X2)
7.	Farm Power (X7)	-0.0759	0.0490	-0.1249	-0.3310(X8), 0.1432(X3), -0.1134(X4), 0.0939(X6), 0.0634(X1).
8.	Total Income (X8)	-0.2442	-0.4676	0.2234	0.1611(X3), -0.1479(X4), 0.1012(X6), 0.0617(X1).

the variable however, moderately substantial and positive which indicate that age on its own is positively related to shouldering of work load which conform to the relationship observed in case of their male counterparts as reported in Table 4.16. The reason why age showed negative correlation coefficient was that indirect effects channelled by variables like 'total income' and 'family education' were negative and so high that they offset the positive direct effect of this variable. The reason of channellising negative direct effect by these two variable would be subsequently discussed in this section of the book.

Caste

The variable caste also yielded negative correlation coefficient indicating thereby that lower caste women undertake heavier work load. But direct effect of this variable does not substantiate this contention as indicated by substantial indirect effect channelled by this variable. Women coming from family of higher income and higher education undertake lighter load of farm work and somehow such families exist mostly in higher caste. Thus, it is not the caste per say but the economic and educational status of the family which is responsible for negative correlation coefficient. One can however, infer that women of lower caste having lower total income and less family education are prone to carry heavier work load of farm work than the women of higher caste who are destined with greater income and higher education. Similar findings were reported by Singh (1968), Awasthy (1982), Gulati (1984), Devi and Reddy (1984) and Kaur (1987) that participation of lower caste women in agricultural activities was more as compared to higher castes.

Family education and total income

Not only that these two variables channelled substantial indirect effect through caste but they also yielded substantial negative direct effects. In families having several educated members, quite a few are in government and semi-government jobs. This is why their contribution in agriculture was of lower order. In order to compensate for their liability to work on farm, many of them had employed permanent labour and the others were found to depend on hired labours. Because of the deployment of larger labour force, women in these families used to work less on the farm which reduced their work load as compared to the work load undertaken by the women of family with lower education. This is the probable reason that with the increase in family education the corresponding decrease in farm work load was found to exist.

The families having higher total income used to keep servants/permanent labour to work on farm. Women members in such families were also found to work less on farm as compared to the women from low income families. The reason is that with the increase in family income, there was reduction in farm work load undertaken by women members of family.

At this point, it appears quite meaningful to refer to the data of total income reported in Table 4.16. In case of men the direct effect of total income was not found substantial but positive unlike direct effect of this variable in case of women. This clearly reveals that employing permanent labour does not reduce the work load of men but reduces the work load of women in case of higher income group. The menfolk in this group work on farm along with permanent labour but do not allow women of the families to work on the farm.

This, therefore, points out the social value attached to the farm work which the respondents hold in their mind. Working of family on farm is valued lower and hence farm family which can afford to substitute the women labour force by employing permanent labour or hiring more labours do not allow women to work on farm. It is only because of economic compulsions that the farmers of lower income allow their wives to work on the farm despite assigning lower social value to this. Similar conclusions were reported by Sinha (1980) and Kaur (1987), also.

Land-holding

Land holding is another variable which has yielded interesting results. In this case also the correlation coefficient was found to be non-significant and negative, whereas direct effect of this variable was moderately substantial and positive. The results, therefore, suggest that land holding on its own, is a function of work load undertaken by the members of the family in cultivation of wheat and paddy. With the increase in land holding the areas under wheat and paddy also increases which would result in increase of work load which fully justifies the positive substantial effect of land holding on work load. The reason why correlation coefficient was found to be negative was

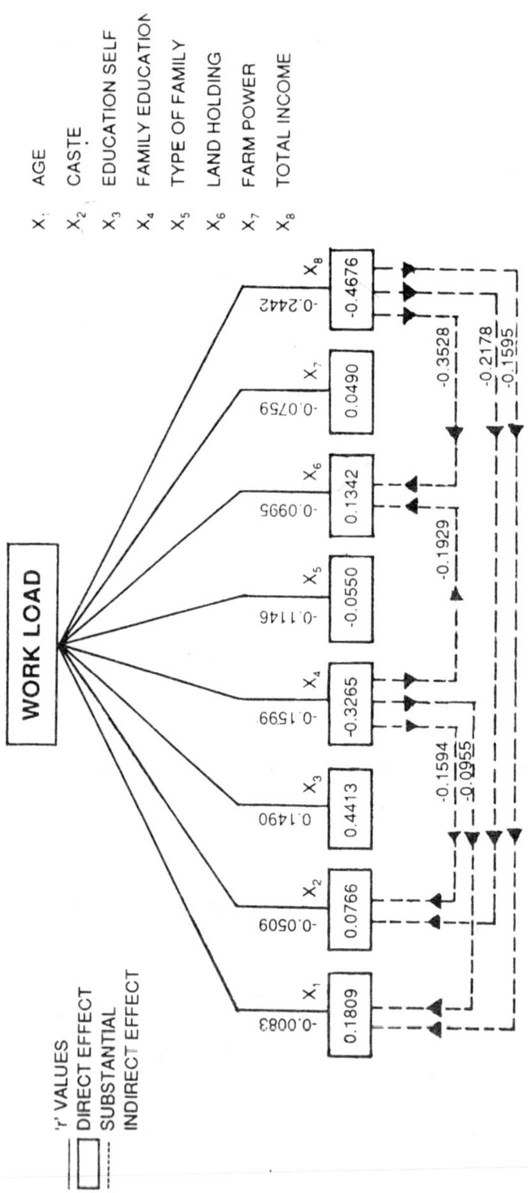

FIG. 4.2: Correlates of work load of women for cultivation of major crops - wheat and paddy

that substantial negative indirect effects were channelled by variables like total income and family education which cancelled the positive direct effect. Land holding was obviously found to be positive and related to family education and total income. In other words, farmers with larger land holding had more educated members, some of whom were having secondary occupations (service etc.) resulting in greater income for the family and thus have employed permanent labour. A cumulative effect of all these has resulted in negative correlation coefficient. Similar findings were also reported by Sinha (1980) and Kaur (1987)

Correlates of Work Load of Men and Women for Cultivation of Major Crops — Wheat and Paddy

Further, the data for men and women were pooled and reset for analysis. The results are reported in Table 4.18. The data reported in Table 4.18 do not provide any significant information. The variables like land holding, total income and farm power yielded substantial direct effects. The other variables did not have substantial direct effect because of the direct effects yielded by the men and women respondents independently were in opposite directions cancelling each other.

Correlates of Work Load of Men and Women for Cultivation of Major Crops — Wheat, Cotton and Bajra

In case of farmers from Hisar district of Haryana State, the data were collected for cultivation of three crops — wheat, cotton and *Bajra*. The work load data of three crops were pooled and its relationship with nine identified correlates were studied for the men, the women and the total respondents. The results are reported in Tables 4.19, 4.20 and 4.21. The results maintain the same trend as reported in case of the respondents of Karnal district for cultivation of wheat and paddy with only a little difference. As reported in Table 4.19, Correlation coefficients of three variables caste, type of family and land holding were found to be significantly and negatively correlated with work load as against none of the variables having yielded significant correlation in case of data collected from Karnal district. The three variables have not only significant correlation but also substantial direct effects. The results of these three variables are discussed as under:

TABLE 4.18: Correlates of Work Load of Men and Women for Cultivation of Major Crops - Wheat and Paddy

Sl. No.	Independent variables	'r' values	Direct effect	Total indirect effect	Substantial Indirect effect through
1.	Age (X1)	0.0059	-0.0003	0.0061	-0.0700(X9), 0.0537(X7).
2.	Caste (X2)	0.0339	0.0373	-0.0034	-0.0957(X9), 0.0775(X7), 0.0676(X8).
3.	Self Education (X3)	-0.0721	-0.0497	-0.0224	-0.0637(X9).
4.	Family Education (X4)	-0.0620	-0.0356	-0.0264	-0.0940(X9), 0.0515(X7).
5.	Social Participation (X5)	-0.0339	-0.0456	0.0117	-Nil-
6.	Type of Family (X6)	-0.0691	-0.0361	-0.0330	-0.0883(X9).
7.	Land holding (X7)	0.0340	0.1030	-0.0690	-0.1373(X9), 0.0885(X8).
8.	Farm Power (X8)	0.0433	0.1347	-0.0914	-0.1385(X9), 0.0677(X7), -0.0052(X5).
9.	Total Income (X9)	-0.0717	-0.2039	0.1332	0.0914(X8), 0.0694(X7).

TABLE 4.19: Correlates of Work Load of Men for Cultivation of Major Crops - Wheat, Bajra and Cotton

Sl. No.	Independent variables	'r' values	Direct effect	Total indirect effect	Substantial Indirect effect through
1.	Age (X1)	-0.2011	-0.0541	-0.1470	0.3236(X4), -0.1658(X6), -0.1079(X3), -0.1025(X2), -0.0542(X7).
2.	Caste (X2)	-0.3199*	-0.2191	-0.1008	0.3430(X4), -0.1790(X3), -0.1343(X6), -0.0685(X7), -0.0514(X5).
3.	Self Education (X3)	-0.1910	-0.3078	0.1168	0.4296(X4), -0.1274(X2), -0.0740(X6), -0.0518(X5).
4.	Family Education (X4)	-0.1267	0.5484	-0.6752	-0.2411(X3), -0.1631(X6), -0.1370(X2), -0.0606(X7).
5.	Social Participation (X5)	-0.2355	-0.1191	-0.1164	0.2250(X4), -0.1339(X3), -0.0946(X2), -0.0529(X7).
6.	Type of Family (X6)	-0.2852*	-0.3102	0.0250	0.2884(X4), -0.0948(X2), -0.0734(X3).
7.	Land holding (X7)	-0.3276*	-0.0860	-0.2416	0.3867(X4), -0.1786(X6), -0.1748(X2), -0.1672(X3), -0.0733(X5).
8.	Farm Power (X8)	-0.2588	-0.0689	-0.1899	0.2230(X4), -0.1307(X6), -0.0916(X3), -0.0835(X2), -0.0590(X4), -0.0575(X7).
9.	Total Income (X9)	-0.2676	0.0555	-0.3231	0.3511(X4), -0.1620(X2), -0.1604(X6), -0.1475(X3), -0.0753(X5), -0.0709(X7).

* Significant at 0.05 level of probability

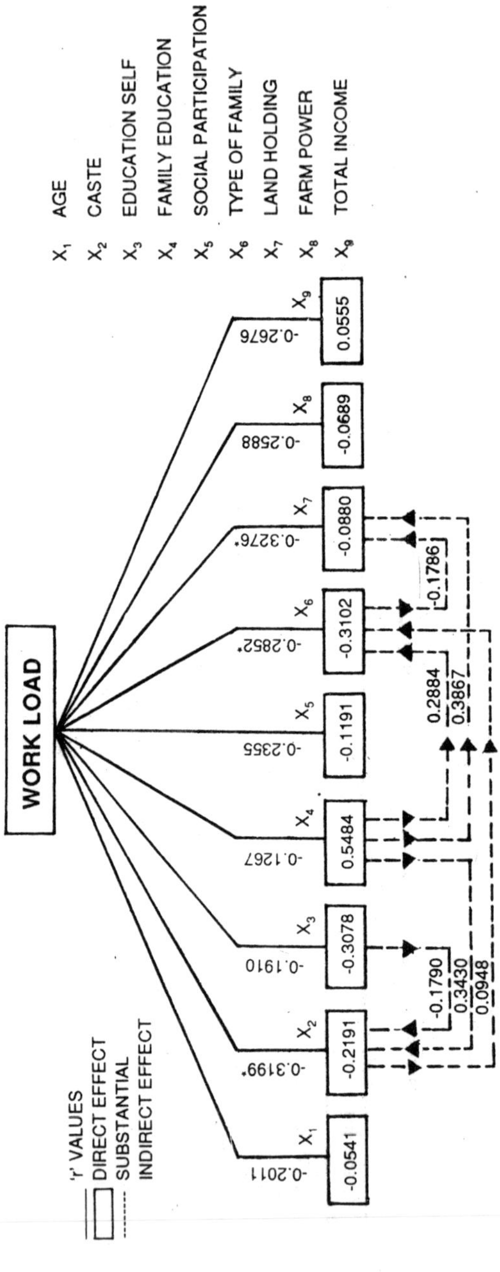

FIG. 4.3: Correlates of work load of men for cultivation of major crops-wheat, cotton and bajra

TABLE 4.20: Correlates of Work Load of Women for Cultivation of Major Crops - Wheat, Bajra and Cotton

Sl. No.	Independent variables	'r' values	Direct effect	Total indirect effect	Substantial Indirect effect through
1.	Age (X1)	0.1213	0.0021	0.1192	0.3472(X7), -0.2202(X2), 0.1886(X6), -0.1045(X8), -0.0777(X4).
2.	Caste (X2)	0.0348	-0.4508	0.4856	0.4211(X7), 0.2235(X6), -0.1182(X8), -0.0821(X4), 0.0598(X9).
3.	Self Education (X3)	0.0410	-0.0383	0.0793	0.2563(X7), -0.1293(X2), -0.1285(X6), -0.0806(X8), -0.0904(X4), -0.0557(X5).
4.	Family Education (X4)	0.0463	-0.1270	0.1733	0.3869(X7), -0.2914(X2), 0.2274(X6), -0.1340(X8), 0.0511(X9).
5.	Social Participation (X5)	-0.0432	-0.1171	0.0739	0.1491(X6), 0.1429(X7), -0.1355(X8).
6.	Type of Family (X6)	0.2114	0.3830	-0.1716	0.2990(X7), -0.2630(X2), -0.1240(X8), -0.0753(X4).
7.	Land holding (X7)	0.0900	0.5267	-0.4367	-0.3604(X2), -0.2177(X8), 0.2174(X6), -0.0932(X4), 0.0664(X9).
8.	Farm Power (X8)	-0.0571	-0.3248	0.2677	0.3531(X7), -0.1640(X2), 0.1463(X60, -0.0524(X4).
9.	Total Income (X9)	0.1215	0.0803	0.0412	0.4355(X7), -0.3360(X2), 0.2351(X6), -0.1714(X8), -0.0807(X4).

TABLE 4.21: Correlates of Work Load of Men and Women for Cultivation of Major Crops - Wheat Bajra and Cotton

Sl. No.	Independent variables	'r' values	Direct effect	Total indirect effect	Substantial Indirect effect through
1.	Age (X1)	-0.1026	-0.1230	0.0204	0.1713(X4), -0.1359(X2), 0.1248(X7), -0.0866(X3), 0.0613(X9), -0.0594(X8).
2.	Caste (X2)	-0.1085	-0.2965	0.1880	0.1839(X4), 0.1576(X7), 0.1045(X9), -0.0934(X3), -0.0660(X8), -0.0564(X1).
3.	Self Education (X3)	-0.1439	-0.2282	0.0843	0.2017(X4), -0.1213(X2), 0.0905(X7), 0.0597(X9), 0.0528(X5).
4.	Family Education (X4)	-0.0207	0.2945	-0.3152	-0.1851(X2), -0.1563(X3), 0.1420(X7), 0.0895(X9), -0.0742(X8), -0.0715(X1).
5.	Social Participation (X5)	-0.1686	-0.1286	-0.0400	0.1068(X4), -0.0937(X3), 0.0857(X7), -0.0807(X8), -0.0706(X2), 0.0587(X9).
6.	Type of Family (X6)	-0.0338	-0.0235	-0.0102	0.1615(X4), -0.1458(X2), 0.1122(X7), 0.0788(X9), -0.0736(X8) -0.0616(X1), -0.0544(X3).
7.	Land holding (X7)	-0.0809	0.1997	-0.2807	-0.2340(X2), 0.2094(X4), -0.1242(X8), 0.1169(X9), -0.1034(X3), -0.7688(X1), -0.0552(X5).
8.	Farm Power (X8)	-0.1267	-0.1892	0.0625	0.1312(X7), 0.1156(X4), -0.1035(X2), 0.0714(X9), -0.0548(X4).
9.	Total Income (X9)	-0.0346	0.1429	-0.1775	-0.2169(X2), 0.1847(X4), 0.1635(X7), -0.0954(X3), -0.0946(X8), -0.0528(X5), -0.0527(X1).

Variables caste as well as land holding yielded significant negative correlation coefficient and also substantial negative direct effect indicating thereby that higher the caste and higher the land holding lower is the per capita work load of the male members of the family. Interestingly, the inter correlation coefficient between caste and land holding was found to be positive and significant (r=0.7396p > 0.01). This means farmers of higher caste were having larger land holding. The researcher's experience and also the collected data suggest that big farmers in this area generally cultivate a small part of their holdings and the larger part is leased out to other farmers of the village. Their substantial earning was from leasing of their lands on hire rather than from crop cultivation. Moreover, because of their being economically better off, they also maintain some permanent labour or depend rather heavily on hired labour. This is the probable reason that their farm work load is relatively lower than those who have smaller land holding and incidently majority of them are from lower castes.

In case of type of family, our data suggest that a great majority of nuclear families hail from lower socio-economic status and they are low land holders (48%). Because of their low land holding they have to work hard on farm and they cannot afford to employ hired labour hence their work load is higher.

EXTENT OF EXPOSURE OF MODERN CROP TECHNOLOGY TO FARM WOMEN

In order to test the knowledge of modern cultivation technology held by farm women, a knowledge test was developed for cultivation of four major crops. The scales contained 14 test items for wheat, 17 test items for paddy, 15 test items for cotton and 16 test items for *Bajra* cultivation. Number of items also indicate the maximum obtainable knowledge score for cultivation of a particular crop. The women respondents on an average were found to hold about 70 per cent knowledge in case of wheat cultivation, 56 per cent in paddy and cotton cultivation and about 61 per cent in case of *Bajra* cultivation. The inter socio-economic status variation in holding this knowledge was not found to be very substantial except in case of paddy in which knowledge gap in high and medium socio-economic status group respondents were found to be relatively substantial. The data clearly indicate that the farm women were so much involved in farming, but they do not hold adequate knowledge of modern crop cultivation technology. Whatever, they have learnt about modern technology is by way of their involvement in farming. There had been no deliberate attempt made to give them education about the modern technology.

However, their knowledge of technology related to the tasks they perform was found to be quite high only in case of paddy cultivation. In case of other crops it was found to be at the medium level. The inter socio-economic strata variation were also not found to be substantial. In this case, the operation which women performed, in fact, do not involve much sophisticated technology except chemical weed control and yet their knowledge level being so low speaks of the magnitude of their ignorance of modern cultivation technology. Thus, there is an obvious need to provide planned exposure of modern farm technology to the women who are so heavily involved in cultivation of crops to set the pace of agricultural development in the area.

TABLE 4.22: Extent of Exposure of Modern crop Technology to Farm Women

Sl. No.	Crop	Activities performed by Women (in percentage)	Total Knowledge (in percentage)
1.	**Wheat**		
	High SES	4.16(69.33)	10.00(71.43)
	Medium SES	4.00(66.67)	9.41(67.21)
	Low SES	4.23(70.50)	10.09(72.07)
	Total	4.15(69.17)	9.88(70.57)
2.	**Paddy**		
	High SES	8.20(91.11)	10.90(64.12)
	Medium SES	7.86(87.33)	8.21(48.29)
	Low SES	7.54(83.78)	9.04(53.18)
	Total	7.76(86.22)	9.56(56.23)
3.	**Cotton**		
	High SES	4.25(60.71)	8.25(55.00)
	Medium SES	3.40(48.57)	7.80(52.00)
	Low SES	4.44(63.43)	9.00(60.00)
	Total	4.10(58.57)	8.52(56.80)
4.	**Bajra**		
	High SES	3.75(53.57)	9.00(56.25)
	Medium SES	4.07(58.14)	9.60(60.00)
	Low SES	4.44(63.43)	10.04(62.75)
	Total	4.22(60.28)	9.74(60.94)

SES (Socio-economic status)

Figure in parentheses indicate percentages of knowledge held out of the maximum obtainable knowledge score.

Comparative Drudgery Index of Farm Men and Women for Different Farm Operations

While introducing the research issues in the first chapter of book some tentative generalisations derived from past studies and writings of the subject were made. They were:

(a) Women's involvement in agricultural tasks by and large is a source of drudgery; and the drudgery ridden agricultural tasks are assigned to women only because of male dominated social order. In order to test the validity of these generalisations, drudgery associated with the major farm operations involved in cultivation of wheat and paddy crops were worked out.

The drudgery scale values of the farm operations calculated by case III Model of Thurstone's law of comparative judgment as discussed in detail in the chapter on research methodology for male and female respondents are reported in Table 4.23.

TABLE 4.23: Drudgery Assigned to the Selected Farm Operations by Men and Women Farm Workers

Sl.No	Operations	Drudgery Assigned By					
		Men	Rank	Women	Rank	Pooled	Rank
1.	Pesticides dusting	0.899	I	1.219	I	0.937	I
2.	Harvesting	0.885	II	0.837	III	0.873	II
3.	Carrying load on head	0.763	III	1.036	II	0.837	III
4.	Spade work during field irrigation	0.673	IV	0.834	IV	0.708	IV
5.	Transplanting	0.363	VIII	0.686	VII	0.538	V
6.	Weeding by *'Kasola'*	0.600	V	0.828	V	0.481	VI
7.	Weeding by *'Khurpi'*	0.501	VI	0.750	VI	0.424	VII
8.	Sowing behind plough	0.394	VII	0.358	VIII	0.214	VIII
9.	Bullock ploughing	0.234	IX	0.000	X	0.148	IX
10.	Pudding by plough	0.000	X	0.138	IX	0.000	X
	Rank order correlation		0.96*				

* Significant at 0.05 level of probability.

The scale values in case of men respondents vary between 0.00 and 0.899 whereas in case of women respondents they vary between 0.00 and 1.219. The scale values of the practices were ranked in descending order of magnitude. As evident from the data (Table 4.23) there is some difference in the two sets of ranks assigned to the practices. The rank order correlation coefficient between them were worked out to be 0.96 which is highly significant. The results, therefore, indicate that the two sets of values obtained from the responses of men and women farm workers are almost alike or atleast very close to each other. It was, therefore, decided to pool the data to work out unified values of the ten practices on the drudgery scale. The values so calculated (again case III Model) are reported in the last two columns of Table 4.23.

As evident from the unified drudgery index, pesticide dusting is considered to have the highest amount of drudgery followed by harvesting, carrying load on head, spade work during field irrigation, transplanting, weeding by *kasola*, weeding by *khurpi*, sowing behind plough, bullock ploughing and pudding by plough.

A further probe in the matter revealed why drudgery was perceived in the selected operations. This is being reported in tabular form as under:

	Operations		Reasons
1.	Pesticide dusting	(a)	Health hazards due to pesticide inhalation.
		(b)	Movement in the standing crop with care not to damage the plants.
		(c)	Discomfort in carrying load of the machine and pesticide.
2.	Harvesting	(a)	Posture of bending or sitting on toe.
		(b)	Energy required in holding a bunch of plants by one hand and cutting them by the other hand.
		(c)	Setting the lodged plants in order and taking care that grains do not shatter.
		(d)	Keeping the harvested plants with adequate care so that bundle making is easy.
3.	Carrying load on head	(a)	Carrying head load of harvested crop with strained movement of eyes and neck.
4.	Spade work during irrigation	(a)	Energy required in the earth work.
		(b)	Discomfort in movement in the cropped field and wet soil.

Contd.

Contd.

	Operations		Reasons
5.	Weeding by *Khurpi**	(a)	Sitting on toe for long hours.
		(b)	Care to avoid injury to cultivated plants while being in thick of them.
6.	Weeding by *Kasola***	(a)	Posture of bending.
		(b)	Energy and care required to uproot the weeds without damaging cultivated plants
7.	Transplanting	(a)	Posture of bending for long hours in standing water.
8.	Sowing behind plough	(a)	Sowing seeds with bent spinal cord and neck.
		(b)	Care and skill in putting seeds uniformly.
9.	Bullock ploughing	(a)	Controlling bullock
		(b)	Obstructed movement in the field
10.	Puddling by plough	(a)	Discomfort in movement in wet field with puddler.
		(b)	Controlling bullocks.

Drudgery Undertaken by Men and Women

Men and women perform various tasks involved in cultivation of wheat and paddy. Data were collected from the selected respondents about their involvement in various operations in terms of man days (8 hours) deployed for one acre of wheat/rice cultivation. The data so collected are reported in Table 4.24 (a & b).

The data reported in Table 4.24(a) indicate that in case of wheat cultivation there are three operations namely pesticide dusting, spade work and ploughing, which are performed exclusively by men. The remaining five operations are such which are performed by men and

* *Khurpi* is a hand tool made of iron (plate and bent) with wooden handle used to weed out unwanted plants by sitting on toes (Fig. 4.4 in Appendix IV).

** *Kasola* is a hand tool made of iron (plate but wider than *khurpi*) with a long wooden handle used to weed out unwanted plants while standing in the field (Fig. 4.5 in Appendix IV).

women both. Out of these five operations three namely 'carry head lead', weeding by *Khurpi* and weeding by *Kasola* are predominantly performed by women (over 80%). Even in case of harvesting about 65% work load is taken up by women. The heavy preoccupation of women in these five wheat cultivation operations has resulted in substantially higher labour contribution by them in cultivation of wheat. On an average a woman devotes 13 man days against only 9.94 man days devoted by a man in cultivation of one acre of wheat. The data also reveal that major preoccupation of women labour in wheat cultivation is in weeding. They spend about half of their man days (deployed in wheat cultivation) only in this operation which they carry either with the help of *Khurpi* or *Kasola*. This appears to be an almost exclusive domain of women labour closely followed by 'on head carrying' of the harvested crop. In case of paddy cultivation the same man dominated operations continued to be their exclusive domain whereas transplanting of paddy was found to be the exclusive domain of women. The remaining three operations are performed by both of them but man days devoted by women heavily outnumbered the man days devoted by men. Weeding and carrying harvested crop on head are predominantly (to the extent of 90%) performed by women. Even in case of harvesting about 70 per cent of the operation is performed by women. The heavy preoccupations in these four operations of paddy cultivation has resulted in much higher contribution of labour by women than by men. On an average a woman devotes 17.3 man days as against only 11.5 man days devoted by a man in cultivation of one acre of paddy.

The generalisations made above gets further support from the data reported in Table 4.25. The data clearly bring to light that there are four operations which are performed by only men and no women. They are 'ploughing', 'puddling', 'spade work during field irrigation' and 'pesticide dusting'. On the other hand, there is only one operation which is performed exclusively by women, i.e., 'transplanting of paddy'. But there are three other operations in which 100 per cent of women labour force is engaged though some men labour force also perform these operations. These are weeding by *Kasola* and *Khurpi* and harvesting. Out of the remaining two operations, 'carrying load on head' is performed by 93.5 per cent of women and only 15 per cent of men. The other operation 'sowing behind plough' is performed by 72.5 per cent of men and only 27.5 per cent of women. If we look back to the data reported in Table 4.23, we find that men dominated operations

TABLE 4.24: Farm operation undertaken by men and women in (a) wheat and (b) paddy cultivation in terms of man days

	Operations										Total	Average
(a) Wheat (n = 100) Drudgery value	Pesticide dusting I (0.937)	Harvesting II (0.873)	Carrying load on head III (0.837)	Spade work during field irrigation IV (0.708)	Transplanting V (0.538)	Weeding by *'Kasola'* VI (0.481)	Weeding by *'Khurpi'* VII (0.424)	Sowing behind plough VIII (0.214)	Bullock-ploughing IX (0.148)	Puddling by plough X (0.00)		
Men	(3.70)*	(28.17)	(2.26)	(20.05)	-	(5.73)	(3.52)	(11.62)	(24.95)	-		
	36.81	280	22.5	199.30		57	35	115.5	248		994.11	9.94
	(100)**	(35.04)	(13.93)	(100)		(13.67)	(11.22)	(95.45)	(100)			
Women	(0)	(39.92)	(10.65)	(0)		(27.69)	(21.31)	(0.42)	(0)			
	(0)	519	138.5	0	-	360	277	5.5	0	-	1300.0	13.0
	(0)	(64.96)	(86.02)	(0)		(86.33)	(88.78)	(4.5)	(0)			
Total	36.81	799.0	161.0	199.30	-	417.0	312.0	121.0	248.0	-	2294.11	22.94
(b) Paddy (n=50)												
Men	(3.61)*	(21.94)	(1.39)	(60.87)	(0)		(3.66)		(6.6)	(1.92)		
	20.74	126	8	349.62	0	-	21	-	38	11	574.36	11.5
	(100)**	(31.27)	(8.08)	(100)	(0)		(9.59)		(100)	(100)		
Women	(0)	(31.99)	(10.51)	(0)	(34.63)		(22.86)		(0)	(0)		
	(0)	277	91	0	300	-	198	-	0	0		
	(0)	(68.73)	(91.92)	(0)	(100)		(90.41)		(0)	(0)	866.0	17.3
Total	20.74	403.0	99.0	349.62	300	-	219.0	-	38	11	1440.36	28.8

* Figures in parentheses indicate the percentage of the total man days over all operations depolyed in the particular operation (horizontal data).

** Figures in parentheses indicate the percentage of the total man days deployed in the operation by men and women (vertical data).

TABLE 4.25: Number of men and women taking up various operations

Sl. No.	Operations	Number of Men	Number of Women
1.	Pesticide dusting	80 (100.00)	0 (0.00)
2.	Harvesting	50 (62.50)	80 (100.0)
3.	Carrying load on head	12 (15.00)	75 (93.80)
4.	Spade work during field irrigation	80 (100.00)	2 (2.50)
5.	Transplanting	0 (0.00)	40 (50.00)
6.	Weeding by '*Kasola*'	27 (33.80)	80 (100.00)
7.	Weeding by '*Khurpi*'	27 (33.80)	80 (100.00)
8.	Sowing behind plough	58 (72.50)	22 (27.50)
9.	Puddling by plough	40 (50.00)	0 (0.00)
10.	Bullock ploughing	80 (100.00)	0 (0.00)

Figures in parentheses are the percentages of men and women performing the task.

are at the position I, IV, IX and X on the drudgery index whereas women dominated operations are at positions II, V, VI and VII on the scale. The next level women dominated operations is on position III whereas men dominated operation is at position VIII on the scale.

Since there is distinct operation wise distribution of labour between men and women farm workers and also there is a variation in the operations in terms of the amount of drudgery involved in them, it is obvious that there is inter-sex difference in the drudgery shouldered by them. In order to take account of this difference, the amount of drudgery taken up by every individual involved in one acre of wheat and paddy cultivation was worked out. The average drudgery data for men and women are reported in Table 4.26 which reveal that there is a significant difference in the total drudgery taken up by men and women in cultivation of wheat and paddy both. In both the cases, the drudgery load of women is more than that of men.

It is however, worth noting that the results did not maintain consistency when average drudgery undertaken per man day of work in cultivation of wheat as well as paddy were worked out. Although the '*t*' value in case of both wheat and paddy cultivation were found to be significant, it is of interest to note that in case of wheat cultivation drudgery per man day taken up by women is more than that by men whereas in case of paddy it is in the reverse order. The results thus clearly reveal that on an average more drudgery prone tasks or operations are performed by women in case of wheat cultivation whereas men perform more drudgery ridden tasks in case of paddy cultivation. Yet the total drudgery suffered by women in both the cases were higher (than men). This is due to the reason that women perform more hours of work than men in cultivation of both wheat and paddy.

TABLE 4.26: Drudgery Taken up by Men and Women in Cultivation of One Acre of Wheat and Paddy

Crop	Average drudgery taken up	't' value	Average drudgery per man days	't' value
Wheat	Men 5.357 Women 8.660	8.825**	0.598 0.693	6.763**
Paddy	Men 7.948 Women 11.199	4.396**	0.705 0.662	3.040**

** Significant at 0.01 level of probability.

The results although do not appear to lend support to the generalisation that drudgery ridden agricultural tasks are assigned to women because of the male dominated social order, the fact that women shoulder greater total drudgery, deserves a particular attention. As reported in Tables 4.13 and 4.14, women shoulder almost entire responsibility of household activities and child rearing and also make a substantial contribution in dairying. In spite of all these, women were found to work for significantly more hours (than menfolk), in cultivation of the two major crops–wheat and paddy shouldering greater total drudgery. This clearly points to the anormity of work load being carried out by women in the farming community.

It is obvious from the findings reported above that the tasks in which women are exclusively or predominantly involved are such that they consume more time and account for greater total drudgery in spite of their drudgery value being not so high. It is, therefore, of great interest to know the bases or the reasons of distribution of the tasks amongst men and women. For this, the respondents both (men and women) were asked to suggest reasons why a particular task is only or atleast predominantly performed by men or women. The data so collected are reported in Table 4.27.

As reported earlier there are four tasks which are performed by men alone. The reasons reported by men as well as women for assignment of these tasks to men are:

(a) Requirement of greater physical power and skill in performance of the tasks.
(b) The social value that men are physically more powerful than women and had greater skill at work etc. There was complete agreement between men and women respondents in assigning these reasons.

TABLE 4.27: Reasons for Operation Wise Division of Labour Amongst Men and Women

Sl. No.	Reasons	Frequency (%)	
		Men	Women
	(a) *Men dominated tasks*		
1.	Needs more physical power	80 (100)	80 (00)
2.	Needs control (animal) and have skill	80 (100)	80 (100)
3.	Social values	80 (100)	80 (100)
	(b) *Women dominated tasks*		
4.	Time consuming	60 (75)	-
5.	Unskilled work	80 (100)	80 (100)
6.	Less wages	80 (100)	-
7.	Monotonous	60 (75)	80 (100)
8.	Less physical power	80 (100)	-
9.	Social values	-	80 (100)

Similarly, there are four tasks which were found to be solely or predominantly performed by women. The reasons of these tasks being assigned to women as perceived by men are (a) time consuming and monotonous nature of task (b) Lower requirement of skill and physical power and (c) less wages being paid for performance of these tasks. So far women respondents are concerned they could only see the tasks to be monotonous and less skillful. They could not see the tasks to be time consuming and low paying. They did not verbalise that their tasks need less physical power but by implication (they had said that men's tasks require more physical power), this can be accepted as their response too.

Interestingly, men as well as women completely agree with each other in verbalising social values as the bases of assignment of these tasks to women. The social values which seem to have played role are that women are physically less powerful and they have lower ability to acquire skill and hence they should take up more time consuming, monotonous and less paying tasks. Men may be physically stronger than women in absolute term, but what was ignored is that the amount of power that the tasks require, is also possessed by women.

The findings thus suggest a very serious implication to a technological solution if innovated to emancipate women from the drudgery they undertake in farm operations/tasks. On the face value of the problem one can easily suggest that if suitable tools and techniques are invented to raise efficiency and reduce time and energy consumption in the tasks being predominantly performed by women, their work load and drudgery shall be reduced. But what if that innovation(s) requires higher skill in operation? Perhaps women will ungrudgingly had over the task to men. The primary task, therefore, is to provide education to men and women both to explode the myth of the value that women are *abala* less powerful and have lower ability to acquire skills. An effort to innovate suitable technology for women and education to both men and women have to go hand-in-hand, otherwise the cherished goal of emancipating women of their drudgery in agriculture will remain unachieved.

IMPACT OF MECHANISATION ON FARM WOMEN'S EMPLOYMENT

An account of mechanical innovations adopted by the respondents lends support to the apprehension raised above. Mechanisation adopted by the farmers under study encompasses use of tractor (for ploughing land and transporting farm produce), thresher, seed cum fertiliser drill, sprayer and duster. Tractor was found to be owned by 15 respondents and 25 others used tractor on hire, thresher was owned by 15 respondents and the remaining 85 used it on hire, seed cum fertiliser drill was owned by 2 respondents and 3 others used it on hire. All of them found to use duster and sprayer whenever they used plant protection chemicals. In response to specific questions framed for the purpose, they unanimously reported that all these five machineries and equipments were operated by men alone since they require skill in handling.

They also reported that use of these equipments did relieve women of certain tasks like carrying harvested crops on head load etc. to some extent but this did not result in unemployment of any reckonable magnitude. In fact, women in general, are so over worked that this small relief of work did not even get mentally registered. Perhaps some more such relief would be welcome. Women respondents however, were of the opinion that these machines and equipments were able to reduce the work load of men quite substantially resulting in an increase in their leisure time to be spent on cards and gossip. This finding thus dose not substantiate the generalisation that mechanisation of agricultural operation renders women unemployed and make them suffer due to reduction in their earning.

The conclusion

The findings reported above lead to the following conclusions:

1. The various operations/tasks involved in cultivation of wheat and rice involve varying amounts/magnitudes of drudgery.
2. The farm operations/tasks are sex specific. Some tasks are performed exclusively by men and some other exclusively or predominantly by women. There are only a few tasks which are performed by men and women almost equitably.
3. The woman's involvement in agronomic operations/tasks in cultivation of wheat and rice is significantly higher than that of men in terms of man days of work done.
4. The total amount of drudgery in carrying out agronomic operations of rice and wheat cultivation, shouldered by women is significantly higher than that shouldered by men. They seem to be in direct proportion of the time spent in performing various tasks by men as well as women.
5. The amount of drudgery per man day of work shouldered by women is significantly higher than that shouldered by men in case of wheat cultivation, but it was is the reverse order in case of rice cultivation. By implication this finding disputes the

assumption that drudgery ridden agricultural tasks are assigned to women because of male dominated social order.

6. The tasks assigned to men are those which are considered to demand greater physical power and skill for performance. The tasks assigned to women are those which are considered demanding less physical power, lower skill and more time, they are also monotonous in nature and low paid. The major force influencing inter-sex assignment of tasks is the social value held by men and women alike that women are *abala*, less powerful and have lower ability to acquire and use skills.
7. The farm machineries in use further demonstrated the play of this value. They were all being operated by men and not by women. This resulted in saving of time and energy of men giving them a greater leisure hours to rest, play cards and gossip. This also gave some relief to women by relieving them of the tasks like carrying head load etc but it was quite marginal. This did not result in any unemployment to the womenfolk.

The findings reported above unmistakably brings to light that men are not callous to women. They consciously do not put women to hardships by assigning them the tasks of greater drudgery. In fact men and women both are unfortunate victims of some unrealistic, unreasonable and undesirable social values. An effort to innovate technologies to reduce drudgery involved in the tasks being performed by women is desirable, but it must accompany a well-designed programme of education for both rural men and women to help them unlearn the unreasonable and undesirable norms and values and to help them inculcate in themselves the new values to achieve quality in life. Unless the two go together emancipation of rural women is unconceivable.

SOCIO-ECONOMIC CORRELATES OF DRUDGERY

As discussed earlier in this chapter of book, the respondents selected for the study came from varying socio-economic standings which might influence the drudgery undertaken by them. It was therefore, considered interesting to see whether the socio-economic characteristics of the respondents influence the drudgery undertaken by them and if so then in what way. For this, the data were reset to employ correlation and path analysis to find out how socio-economic characteristics of the respondents effect the drudgery undertaken. The data are reported in Table 4.28 for the two crops under study with the nine socio-economic characteristics identified for the purpose.

The data reported in Table 4.28 reveal that none of the socio-economic characteristic of the respondents yielded any significant correlation coefficient with the drudgery undertaken by men in cultivation of paddy as well as wheat crop. However, some of the direct and indirect effect point to some interesting results which are discussed as under:

Caste

The direct effect of caste on amount of drudgery undertaken was very low in case of the wheat farmers but was quite substantial and positive in case of paddy farmers. This is largely because of the fact that the plant protection measures like dusting and spraying which has the highest drudgery coefficient (0.937) are undertaken in paddy cultivation and not in wheat cultivation. Further, these operations were found to be performed by the land owners themselves and not by the casual labourers since it is believed to involve skill in handling the dusters/sprayers. The data also indicated that higher caste respondents were big land owners whereas a great majority of lower caste respondents were landless or marginal land owners who used to work on other's farms as casual labourer. This seems to be the major reason that the respondents of higher caste were found to undertake high drudgery in case of paddy cultivation (and not in wheat cultivation). It is of further interest to know that even in case of paddy cultivation the direct effect was positive and substantial and correlation coefficient was found to be negative and non-significant. This is largely because of substantial negative direct effects channelled by variables like total land holding and farm power. All these variables were also found to have substantial negative direct effects in case of paddy cultivation which are discussed hereafter.

Land Holding

Land holding was found to yield substantial but negative direct effects and also negative correlation coefficient in case of both wheat and paddy cultivation. There are two possible reasons for reduction in

TABLE 4.28: Socio-economic Correlates of Drudgery Undertaken by Men in Wheat and Paddy Cultivation

Sl. No.	Independent variables	Crop	'r' values	Direct effect	Total indirect effect	Substantial indirect effect through
1.	Age (X1)	Wheat	-0.1387	-0.0469	-0.0918	-0.1626(X7), 0.0673(X9), -0.0646(X4), 0.0637 (X3)
		Paddy	-0.1932	-0.0128	-0.1804	0.1636(X2), -0.1528(X7), 0.1352(X9), -0.1169(X3), -0.1080(X8), -0.0812(X6)
2.	Caste (X2)	Wheat	-0.1242	-0.0605	-0.0637	-0.1799(X7), 0.0917(X3), 0.0850(X9), -0.0715(X4)
		Paddy	-0.1161	0.2904	-0.4065	-0.1953(X7), 0.1794(X9), -0.1435(X8), -0.1367(X3), -0.0718(X6)
3.	Self Education (X3)	Wheat	-0.0366	0.1526	-0.1892	-0.1229(X7), -0.1028(X4), 0.0664(X9)
		Paddy	-0.2715	-0.2293	-0.0422	0.1732(X2), 0.1289(X9), -0.1184(X7), -0.1019(X6), -0.0700(X4), -0.0593(X8)
4.	Family Education (X4)	Wheat	-0.0943	-0.1393	0.0450	-0.1554(X7), 0.1125(X3), 0.1047(X9)
		Paddy	-0.2260	-0.0972	-0.1287	0.1749(X9), -0.1650(X3), 0.1405(X2), -0.0991(X7), -0.0959(X6), -0.0848(X8).
5.	Social Participation (X5)	Wheat	-0.0314	-0.0009	-0.0305	-0.1416(X7), 0.1189(X9)
		Paddy	0.1290	0.2103	-0.0813	Nil
6.	Type of family (X6)	Wheat	-0.0391	0.0561	-0.0952	-0.1254(X7), 0.0848)x9), -0.0732(X4), 0.0574(X3)
		Paddy	-0.2138	-0.1784	-0.0354	0.1648(X9), -0.1309(X3), 0.1169(X2), -0.0853(X8), -0.0796(X7), -0.0523(X4).
7.	Land holding (X7)	Wheat	-0.1656	-0.2684	0.1028	0.1427(X9), -0.0812(X4), 0.0687(X3)
		Paddy	-0.1939	-0.2482	0.0543	0.2284(X2), 0.2175(X9), -0.1711(X8), -0.1094(X3), -0.0572(X6)
8.	Farm Power (X8)	Wheat	-0.0917	0.0221	-0.1138	0.1005(X9)
		Paddy	-0.1441	-0.2817	0.1376	0.2411(X9), -0.1508(X7), 0.1478(X2), -0.0540(X6)
9.	Total income (X9)	Wheat	-0.0468	0.1885	-0.2353	-0.2032(X7), -0.0774(X4), 0.0529(X3)
		Paddy	-0.0357	0.3711	-0.4068	-0.1830(X8), -0.1455(X7), 0.1403(X2), -0.0796(X3), -0.0792(X6)

drudgery with the increase in land holding. One is that with the increase in land holding reliance on hired labourer both on casual and permanent basis were found to increase who shared part of their drudgery. On the other hand the casual labourers are the landless or marginal/small land holders whose drudgery increases because of working on the farms of big land holders. Hence, it operates as a double edged weapon. The other reason is that in case of large land holding operations like ploughing and puddling are carried out by own or hired tractors rather than by bullocks which is exclusively done by men, so to that extent their drudgery is reduced as against farmers of lower holdings who carry out all the operations by bullock driven ploughs.

Farm Power

Farm power yielded no significant correlation coefficient in case of both wheat and paddy cultivation. But interestingly, it yielded almost negligible but positive direct effect in case of wheat cultivation, whereas negative and quite substantial direct effect in case of paddy cultivation. One of the reasons for the differential direct effects may be the nature of cultivation of the two crops. Puddling is one of the drudgery prone operation carried out by men in case of paddy cultivation. This operation in case of those having more farm powers is carried out by tractor and not by bullock which substantially reduces their drudgery as compared to farmers of low farm power who have to carry out this operation with a pair of bullocks. One may agree that the same factor should also reduce drudgery in case of wheat cultivation since land preparation for sowing wheat might be carried out by tractors on the farms of big land holders. For the sake of argument it appears quite logical but the fact is that the wheat cultivation data were collected from farmers of Hisar district where the number of tractors owned by them was much less than those of Karnal district from where the data regarding paddy cultivation was collected. In case of the respondents from Hisar district, higher farm power means more pairs of bullocks and not ownership of tractors.

Total Income

It is quite interesting that in case of both wheat and paddy cultivation, total income of the respondents yielded negative but quite negligible

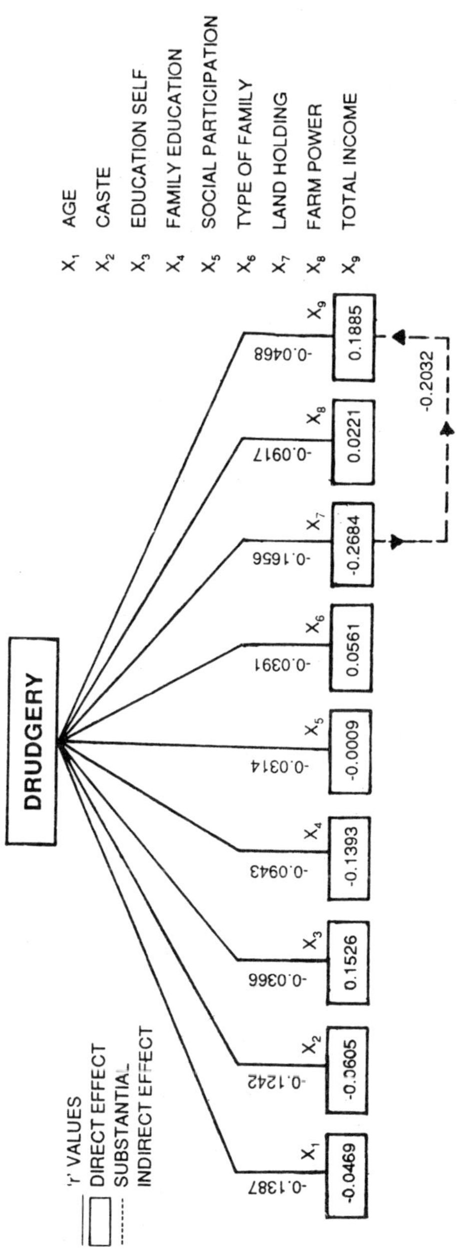

FIG. 4.4: Socio-economic correlates of drudgery undertaken by men in wheat cultivation

correlation coefficient, but they yielded positive and quite substantial direct effects. One of the reasons for substantial positive direct effect was that the farmers with higher income were found to own dusters and sprayers and liberally used pesticides as well as weedicides. Though, weedicides were used in both the crops, pesticides were used only in case of paddy and this is the reason that the direct effect in case of paddy was found to be much more substantial than wheat. This operation alone is capable of affecting the drudgery load since it has the highest drudgery quotient as reported in Table 4.23 (0.937). The reason why the '*r*' values were so greatly reduced was the indirect negative effects channelled by variables like farm power, land holding etc. with which this variable has significant positive correlation.

The data of the relational analysis among socio-economic characteristics and load of drudgery undertaken by women respondents in cultivation of both wheat and paddy crops are presented in Table 4.29.

The data reported in Table 4.29 clearly bring to light that all the correlation coefficients were not only non-significant but were also extremely low in magnitude. The direct effect were also of very low magnitude in case of most of the variables. The direct effect in case of total income was found to be the most substantial (though not very high) in wheat cultivation and somewhat substantial in case of paddy cultivation. In both the cases, they were found to be negative indicating thereby that with increase in income drudgery load of women was reduced in both the crops. It appears logical since in case of family of higher income, employment of casual and even permanent labour is higher and working of family women on farm is restricted as discussed in the section of work load reported in this chapter. This, therefore, reduces the drudgery load of women with increase in family income. Another variable in case of wheat cultivation which yielded moderately substantial direct effect was social participation and in case of paddy cultivation it was self-education. But this data do not make much sense since the social participation level of women from Hisar district (wheat cultivation respondents) was found to be extremely low and a great majority of women respondents from Karnal district (Paddy cultivation respondents) were found to be illiterate with only few having education up to primary level.

It may be summed up from the data that the socio-economic characteristics fail to influence the load of drudgery undertaken by farm women in cultivation of both wheat and paddy crops.

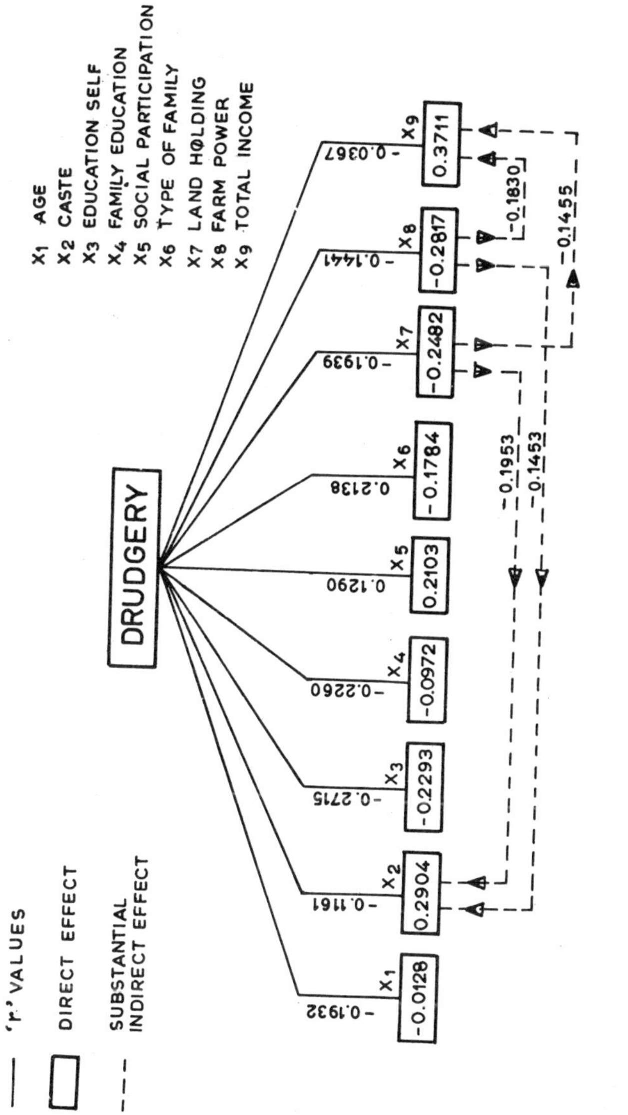

Fig. 6. SOCIO-ECONOMIC CORRELATES OF DRUDGERY UNDERTAKEN BY MEN IN PADDY CULTIVATION

FIG. 4.5: Socio-economic correlates of drudgery undertaken by men in paddy cultivation

TABLE 4.29: Socio-economic Correlates of Drudgery Undertaken by Women in Wheat and Paddy Cultivation

Sl. No.	Independent variables	Crop	'r' values	Direct effect	Total indirect effect	Substantial indirect effect through
1.	Age (X1)	Wheat	0.0288	0.0598	-0.0309	-0.0938(X9), 0.0561(X2)
		Paddy	0.0266	0.2280	-0.2014	-0.1036(X2), 0.0579(X8), -0.0549(X6), -0.0505(X9)
2.	Caste (X2)	Wheat	0.0953	0.1114	-0.0161	-0.1117(X9)
		Paddy	-0.0977	-0.1746	0.0769	0.1353(X1), 0.0818(X8), -0.0690(X9), -0.0524(X6)
3.	Self Education (X3)	Wheat	-0.1001	-0.0060	-0.0941	-0.0982(X9), -0.0705(X5)
		Paddy	0.1500	0.2604	-0.1103	-0.0541(X9), 0.0536(X8)
4.	Family Education (X4)	Wheat	-0.0166	0.0593	-0.0759	-0.1359(X9), 0.0598(X2), -0.0591(X5), 0.0518(X6)
		Paddy	-0.0401	-0.0686	0.0285	0.1300(X3), -0.0852(X2), -0.0672(X9), 0.0667(X1), -0.0638(X6), 0.0574(X8)
5.	Social Participation (X5)	Wheat	-0.2238	-0.2164	-0.0074	Nil
		Paddy	0.0000	0.0000	0.0000	Nil
6.	Type of family (X6)	Wheat	0.0183	0.0955	-0.0772	-0.1159(X9), -0.0563(X5), 0.0535(X2)
		Paddy	-0.0741	-0.1230	0.0488	0.1018(X1), -0.0744(X2), 0.0692(X3), -0.0626(X9), 0.0575(X8)
7.	Land holding (X7)	Wheat	-0.0915	0.0093	-0.1008	-0.1949(X9), 0.0716(X2), -0.0525(X5)
		Paddy	-0.0287	-0.0158	-0.0129	-0.1263(X2), 0.1156(X8), -0.1119(X9), 0.1116(X1) 0.0933(X3), -0.0545(X6)
8.	Farm Power (X8)	Wheat	-0.1480	-0.0591	-0.0889	-0.1364(X9), -0.0704(X5)
		Paddy	0.0604	0.1651	-0.1047	-0.1049(X9), -0.0865(X2), 0.0844(X3), 0.0800(X1)
9.	Total income (X9)	Wheat	-0.1674	-0.2523	0.0848	Nil
		Paddy	-0.0348	-0.1483	0.1135	0.1169(X8), 0.0951(X3), -0.0813(X2), 0.0778(X1), -0.0519(X6)

CONTRIBUTION OF WOMEN IN CULTIVATION OF MAJOR CROPS

As has already been reported, women perform 100 per cent of household activities and about 70 per cent of dairying activities in terms of work hours. In addition, they also spend substantial time and energy in cultivation of crops. In order to make an estimate of their contribution in cultivation of crops, data related to per acre cultivation of major crops of the area were collected in terms of man days (8 hours) deployed by men and wcmen of various land holdings. The data for four major crops are reported in Table 4.30. The landless respondents were excluded for the purpose of this section of report and only the time devoted by family labour (men and women) were taken into account.

The data reported in Table 4.30, make several interesting revelation which are as follows:

TABLE 4.30: Contribution of Women in Cultivation of Major Crops

Sl. No.	Crop	Land holding	No. of respon-dents	Average man days per acre deployed by			
				Men	%	Women	%
1.	Wheat	Upto 2.5 acre	17	17.74	47.70	19.45	52.30
		2.51-5.0 acre	28	15.15	50.18	15.04	49.82
		5.1-7.5 acre	13	9.06	59.49	6.17	40.51
		Above 10 acre	14	3.54	74.53	1.21	25.47
		Total	72	12.52	50.89	12.08	49.11
2.	Paddy	Upto 2.5 acre	11	21.00	51.95	19.43	48.15
		2.51-5.0 acre	12	15.08	61.25	9.54	39.75
		5.1-7.5 acre	8	10.34	83.79	2.00	16.21
		Above 10 acre	4	6.81	90.80	0.69	9.20
		Total	35	14.92	60.12	9.91	39.88
3.	Cotton	Upto 2.5 acre	6	21.21	50.40	20.87	49.60
		2.51-5.0 acre	16	22.87	40.62	33.43	59.38
		5.1-7.5 acre	5	10.10	63.52	5.80	46.48
		Above 10 acre	10	6.55	82.39	1.40	17.61
		Total	37	16.57	45.17	20.11	54.83
4.	Bajra	Upto 2.5 acre	6	19.00	48.67	20.04	51.33
		2.51-5.0 acre	16	18.59	48.51	19.73	51.49
		5.1-7.5 acre	5	9.40	61.44	5.90	39.56
		Above 10 acre	10	2.90	78.38	0.80	21.62
		Total	37	13.32	50.65	12.98	49.35

1. Out of the total family labour force deployed in cultivation of crop almost equal contributions were made by the men and women family labour force incultivation of wheat and *Bajra* crops. In cultivation of paddy about 60 per cent labour force was contributed by men and only 40 per cent by women, whereas in case of cotton only about 55 per cent of labour force by men and 45 per cent by women. The reason why more men labour force was deployed in case of paddy is a number of operations which are exclusively or dominantly carried out by men. Such operations are plant protection measures, irrigation and spade work in bund making which are exclusively performed by men. Moreover such operations are repeatedly carried out in paddy unlike wheat in which almost no plant protection measure is taken and the number of irrigation is limited to about four or five (in case of paddy number of irrigation is about twenty). In addition, men make substantial contribution in carrying harvested crop from field to threshing ground and in threshing the crop manually. In contrast, threshing is done by machine in case of wheat which consumes considerably less work hours. The reason why women labour force was greater than men in case of cotton is the operation of picking which is exclusively done by women. In case of other crops men and women undertake the harvesting jointly. Also labour required for picking cotton from one acre of land is much more than harvesting of one acre of Paddy or Wheat or *Bajra*. Further, this crop is grown almost unirrigated and hence the engagement of men labour force is relatively less in this crop.
2. The per acre deployment of labour force reduces with increase in size of holding consistently in case of cultivation of all the four major crops. This is because of the reason that with increase in land holding reliance on hired labour (casual as well as permanent) increases. Secondly, the menfolk might be working for as many days and hours as the men from lower land holding but the average contribution of labour per acre would naturally reduce with the increase in acreage.
3. The same trend of decrease in man days with increase in size of holding, holds good in case of family women labour force also, but with a difference. The decrease in case of women is sharper than that of men in case of all the four crops. For example in

case of marginal farmers' families (up to 2.5 acres of holding) contribution of men and women labour force are almost equal in all the four crops whereas in case of the big farmers' families (above 10 acres) the ratio of contribution of men and women labour force is approximately 75 : 25 in wheat, 91 : 9 in paddy, 82 : 18 in cotton and 78 : 22 in *Bajra*. This clearly brings to light that in case of poor farming families men and women work in the field alike whereas in case of bigger farm families women are not allowed to work in the field as much as the men work. Thus the social value that "working in field for family women is less dignified", plays its role in case of better off farm families or big land owners' families.

DECISION-MAKING PATTERN IN THE FARM FAMILY

For the purpose of studying decision-making pattern in the farm family, six distinct areas of decisions generally taken in the family were identified. The major decision-making areas identified in the study related to home-management, child rearing, money management, socio-religious rituals, farm activities and dairying. Information from each of the respondent's family whether the decisions in the specific items on the six major areas are taken by husband, by wife or jointly by both of them were collected through a structured interview schedule. The data are presented in Table 4.31.

The data reported in Table 4.31 reveal that there is only one area in which wives are the major decision-maker. There are three areas in which husbands are major decision-maker and there are two areas in which majority of decision taken jointly by husbands and wives.

In case of home management related decisions 52 per cent are taken by wives, 21 per cent by husband and about 27 per cent jointly by husbands and wives. This trend of decision-making was found to prevail almost uniformly in case of respondents of high, medium and low socio-economic categories.

It is generally believed that child rearing is the exclusive domain of wives but this does not seem to be properly understood. Some of the child rearing chores like cooking and serving food, cleaning clothes etc are certainly performed by wives perhaps almost exclusively but they are not allowed to take decisions on their own in matters related to children's growth and development. Only 4 per cent of such decisions were found to be taken by wives independently whereas 39 per cent of

TABLE 4.31: Decision-making Pattern in Farm Families

Sl.	Decisions	High socio-economic status			Medium socio-economic status			Low socio-economic status			Total respondents		
		n = 18			n = 29			n = 53			n = 100		
		Husband	Wife	Joint	Husband	Wife	Joint	Husband	Wife	Joint	Husband	Wife	Joint
1.	Home Management related decisions	19.00 (21.11)	53.11 (58.88)	18.00 (20.00)	31.00 (21.38)	66.00 (45.52)	48.00 (33.10)	55.00 (20.75)	143.00 (53.97)	67.00 (25.28)	105.00 (21.00)	262.00 (52.40)	133.00 (26.60)
2.	Child related decisions	87.00 (48.33)	12.00 (6.67)	81.00 (45.00)	60.00 (20.69)	2.00 (0.69)	228.00 (78.62)	243.00 (45.85)	26.00 (4.91)	261.00 (49.24)	390.00 (39.00)	40.00 (4.00)	570.00 (57.00)
3.	Money related decisions	164.00 (60.74)	16.00 (5.93)	90.00 (33.33)	229.00 (52.64)	16.00 (3.68)	190.00 (43.68)	509.00 (64.03)	70.00 (8.80)	216.00 (27.17)	902.00 (60.13)	102.00 (6.80)	496.00 (33.07)
4.	Socio-religious Rituals related decisions	40.00 (27.78)	7.00 (4.86)	97.00 (67.36)	42.00 (18.10)	3.00 (1.30)	187.00 (80.60)	62.00 (14.62)	19.00 (4.48)	343.00 (80.90)	144.00 (18.00)	29.00 (3.62)	627.00 (78.38)
5.	Farm related decision	153.00 (94.44)	0.00 (0.00)	9.00 (5.56)	225.00 (86.20)	0.00 (0.00)	36.00 (13.80)	406.00 (85.12)	9.00 (1.89)	62.00 (12.99)	784.00 (87.11)	9.00 (1.00)	107.00 (11.89)
6.	Animal related decision	10.00 (55.56)	0.00 (0.00)	8.00 (44.44)	16.00 (55.17)	0.00 (0.00)	13.00 (44.83)	35.00 (66.04)	1.00 (1.89)	17.00 (32.07)	61.00 (61.00)	1.00 (1.00)	38.00 (38.00)

Figures in parentheses indicate percentages.

decisions were taken by husbands alone and 57 per cent by both of them. In case of family of high socio-economic status, maximum number of such decisions (48.33%) were found to be taken by husbands alone whereas 6.7 per cent decisions were taken by wives alone. It is in case of the families of medium socio-economic status respondents that joint decisions heavily outnumbered the individual decisions of both husbands and wives which tilted the balance in favour of joint decisions in case of the total sample. Similar findings were reported by Puri (1968), Rani and Bhave (1981), Bajwa (1984), Ghosh (1987) and Kaur (1987).

The money management related decisions which includes item like purchases for family needs and sales of farm and family produce remain predominant domain of husbands (60.13%) and wives' views were taken into account only in one-third of decisions. Almost similar trend seems to flow through all the socio-economic strata of farm families. A similar finding was also reported by ILO (1975).

Most of the socio-religious rituals are performed jointly by men and women and most of the decisions are also taken by them jointly (78.38%). The extent of decisions taken jointly by husband and wife found to reduce in case of the high socio-economic status families along with corresponding increase in the decisions taken by husbands alone. The trend of decision-making in other socio-economic status group were found to be almost alike. Similar conclusions were arrived at by Kaur (1987) also.

The farm related decisions were found to overwhelmingly rest with husbands with wives' contribution made marginally in case of joint decisions which constituted only 11 per cent of the total decisions made in this area. The independent decisions taken by wives are negligible (only 1%). The trend is almost uniform throughout the socio-economic strata of families. As discussed earlier in this section of book, farm related activities are performed almost equally by the womenfolk but their contribution in decision related to this was found to be so meagre. Somehow the cultural value that 'farm work is the domain of men and household the domain of women' prevail in the society without any regard paid to the contribution made by women in farming. This value obviously influences the decisions about farm related activities taken in the farmers' families. In case of home management related activities (household activities) men seems to poke their nose and about 27 per cent of the decisions are taken by them jointly with women and other 21 per cent independently but in case of farm related decisions they allow women to take part in the decisions to the extent of only 12 per cent and take independent decision to the extent of only 7 per cent. Domination of males in farm decisions has also been revealed by the studies conducted by Puri (1972), Sharma and Singh (1970), Mulay (1974) Nagre (1976), Jhansi (1981), Yadav (1982) and Kaur (1987) at different points of time and cultural contexts.

Even in case of dairying related decisions in which about 70 per cent of work load is shouldered by women, 61 per cent decisions are taken by husband independently and the remaining decisions taken jointly by both of them. Independent decisions by women is only negligible one per cent. A similar finding was also reported by Kaur (1987).

The overall picture emerges from the extent of decisions taken in farm family in all the six areas of decision-making is that men heavily predominate the decision-making processes and women are denied of their due share.

The cultural values which has provided dominating position to men in the families, have obviously influenced the decision-making process as discussed above. Yet interestingly, there is no sign of dissent from womenfolk against the domination of men. This is well-accepted and there is complete harmony on this in the families.

For the sake of harmony, such values are welcome, but they have also cultivated a sense of inferiority in the womenfolk, which is quite undesirable. This is discernible in the value they attach to the tasks they perform. In response to a question 'how much prestige they attach to the different household tasks performed in the family', the women respondents reported their prestige ratings which are presented in Table 4.32, along with the levels of performance of the task by men and women independently as well as jointly.

It may be mentioned that the prestige was scored on a five point continuum ranging from one to five, the median value being three. The data reported in Table 4.32 reveal that the highest prestigious score was assigned to the socio-religious tasks which are performed predominantly by husband and wife jointly (81%). The next prestigious tasks (which is only slightly lower) are money management related tasks which are predominantly performed by men (79%). On the other hand, two tasks most predominantly performed by women (98.93% and 97.44%) were given prestige score of less than two out of five which is much below median score (food and household related tasks).

The child related tasks which are performed by women to the extent of about 80 per cent and jointly with men to the extent for about 18 per cent could receive prestige score very close to median value of the scale which may be termed as neither high nor low. The data thus clearly point to the fact that in the eyes of the women the tasks which they perform are not prestigious and the tasks which their husbands perform or they perform with their husbands and not independently are significantly more prestigious. This speaks of the sense of inferiority the womenfolk suffers from in our society. There is a need to pay attention to this so that the sense of inferiority goes from them without disturbing the harmony in the family. This is important for the society to grow and track on the path of development.

TABLE 4.32: Prestige Attached to Various Household Tasks

Sl. No.	Tasks	Prestige score	Performed by		
			Men (%)	Women (%)	Joint (%)
1.	Food related tasks	1.70	0.86	98.93	0.21
2.	Household related tasks	1.93	1.00	97.44	1.56
3.	Child related tasks	3.54	0.00	81.89	18.11
4.	Money related tasks	4.82	79.00	2.25	18.75
5.	Socio-religious tasks	4.99	12.50	6.25	81.25
	Overall	3.02	9.68	75.82	14.50

UTILISATION PATTERN OF SAVINGS

It was found that 65 of the 100 respondents were able to make some savings from their annual earnings in the immediate preceding year. They were asked about the use they could make of the savings. The following responses were obtained as reported in Table 4.33.

It is gratifying to note that about half of them (31) had some savings thrift which they exhibited by making bank deposits of their savings, twenty-two of them made some asset out of the savings by way of making investments in construction of house, purchase of vehicle, agricultural implements and buffalo. Four of them spent on ornaments whereas five of them spent their savings on social obligations like marriage of children. Three of them spent on children's education and health which by any standard is small for the most valuable investment made for the future of family in the society.

TABLE 4.33: Utilisation Pattern of Savings

Sl. No.	Item	Number	Percentage
1.	*Asset Building*		
	(a) House construction	15	23.08
	(b) Vehicle	1	1.54
	(c) Agricultural implements	3	4.61
	(d) Buffalo	3	4.61
	Total	22	33.84
2.	Bank deposits	31	47.69
3.	Ornaments	4	6.16
4.	Marriage of children	5	7.69
5.	Children's education	1	1.54
6.	Children's health	2	3.08
	Total	65	100.00

5

SUMMARY AND CONCLUSION

Women are almost half of the adult population. They constitute one-third of the labour force but consume two-thirds of the world's working hours and yet earn only one-tenth of the income and own only one per cent of the world property (United Nations, 1975).

The prosperity and the growth of a nation depend on the status and development of its women as they not only constitute nearly half of its population. If a total range of women's actual economic activities were to be taken into account, all over India, women work long hours in market and non-market economy, and more so in rural India. This draws our attention to the plight and drudgery of rural women. The State of Haryana where this study was conducted is no exception to this. The major issues which the present investigation proposed to address itself, are the following:–

1. The multiple role of women in agriculture, allied occupations and household activities are generally, underestimated and undervalued. By and large, they have remained as "invisible hands".
2. The nature and extent of women's involvement in agricultural and non-agricultural activities vary greatly from region-to-region and even within a region, it varies widely among different farming systems and socio-economic groups of families.
3. Women toil hard in labour intensive jobs but have little access to the knowledge of modern farm technology.
4. Women's involvement in agricultural tasks by and large is a source of heavy burden of drudgery on them. Drudgery ridden agricultural tasks are assigned to women probably because of

male dominated social order.

5. Whenever technological innovation(s) to reduce drudgery on a particular task, is available, it results in displacement of women from the tasks sometimes rendering them unemployed (partially of fully).

There had been no attempt to test these assumptions empirically. The present investigation was therefore, designed to test the tenability of these assumptions through empirical analysis of data with the following specific objectives:–

1. To study the nature and extent of involvement of rural women and men of various socio-economic strata in agriculture and home activities in terms of operations, time spent, responsibility and accountability.
2. To study the extent of exposure of modern crop technology to farm women.
3. To prepare a comparative drudgery index for agricultural operations generally taken up by/assigned to women and men in the community/locality.
4. To study the impact of mechanisation on farm women especially in terms of employment, displacement, work load and income.
5. To study the extent of contribution of women in cultivation of major crops.
6. To study the role of women in decision-making in farming, home and other activities.
7. To study the utilisation pattern of savings in the family.

The study was conducted in Haryana State. Haryana State has been divided into three zones, i.e., hot-aid, hot-semi-arid and hot-humid zone on the basis of agro-climatic conditions. Two districts namely Karnal and Hisar from two regions, i.e., hot-humid and hot-semi-arid, respectively of the State were purposively selected for this study. Two villages from each district, i.e., Mehmoodpur and Gogripur of Karnal and Shahpur and Patan of Hisar district were selected randomly for collection of data. Twenty-five men and women from each of the four selected villages representing high, medium and low socio-economic strata of the village community were randomly selected for collecting data for the present investigation. The data were collected with the help of a structured interview schedule. Ten most important practices of wheat and paddy cultivation were selected to prepare a drudgery index. A semi-projective type instrument was developed to collect respondent's perception of drudgery involved in the ten selected practices. This data were collected from 30 farm women and men (their husbands). The data thus collected were subjected to different statistical analyses. The major findings of which are summarised below:

Socio-economic Profile

Age of the respondents varied from 20 to 55 years with an average of 35 years. Majority of women from high and medium socio-economic strata were of middle age group and those of low socio-economic stratum were relatively younger. Most of the respondents of high and medium socio-economic status groups belonged to upper caste and those of low socio-economic stratum belonged to lower caste. However, most of the women of high socio-economic stratum had obtained formal education up to middle class whereas a great majority of the women from low socio-economic stratum were illiterate in both the district. About half of the medium socio-economic stratum women had formal schooling up to middle and matriculation standard in Karnal district. However, in Hisar district, a thumping majority of women from medium socio-economic stratum were illiterate. The educational level of husbands, coming form high and medium socio-economic status were significantly higher than those coming form lower socio-economic stratum. Most of the respondents of high socio-economic stratum had higher level of family education than medium and low socio-economic strata respondents. It is, therefore, obvious that family education of high socio-economic group had higher level of schooling may be because they could afford the expenditure involved.

Farm women generally had low social participation. In Hisar district, about one-fourth of the respondents of high and one eighth of medium socio-economic strata were found to participate in village *Mahila Mandals* and the participation of the women of low socio-economic stratum was found to be almost negligible. In Karnal district, about one-tenth of the respondents of high socio-economic stratum were found to participate in village *Mahila Mandals* and the participation of the women of medium low socio-economic strata were almost negligible. Majority of the respondents of high socio-economic stratum had joint family system. The number of joint families were found to reduce with reduction in the socio-economic status of the family. Cultivation was the main occupation of the respondents of high and medium socio-economic strata, whereas, majority of the respondents of low socio-economic stratum were agricultural labourers. All except one of the respondents of high, half of the medium and about one-fourth of low socio-economic strata had *pucca* houses. However, other respondents of medium socio-economic stratum had mixed type of houses and *Katcha* houses were owned by the respondents of low socio-economic stratum only.

The average size of land holding of the respondents of high, medium and low socio-economic status groups were 40.30, 5.43 and 1.40 acres, respectively. Almost all the respondents of high and medium socio-economic status groups had one or two draft animals. However, half of the landless respondents also kept either one or two draft animals because they had bullock-cart for hiring out. Most of the respondents of high socio-economic stratum were having tractors along with the draft animals. Respondents of high socio-economic status group were equipped with improved implements and machineries also. However, the respondents of medium and low socio-economic status group possessed only some traditional or bullock driven agricultural implements. On an average, respondents form high, medium and low socio-economic status possessed about 3, 1 and less than 1 milch animals respectively.

Possession of household appliances by the respondents of high socio-economic stratum were found to be of worth Rs. 1,467 against worth Rs. 1,082 and merely Rs. 559 by medium and low socio-economic status groups. The prestige items were possessed mostly by the respondents of high socio-economic stratum. The average possession came out to be worth over Rs. 19,000 as compare to Rs. 1,300 and Rs. 14 only by the respondents of medium and low socio-economic status groups. The average family income of the respondents drawn from high socio-economic stratum came out to be Rs. 1.19 lakh followed by Rs. 33.5 thousands by medium and Rs. 14.5 thousands by low socio-economic status group respondents. In terms of the percentage of total earning, the expenditure on maintenance (food, cloth and medicines) was found to be the highest for the respondents of low socio-economic status group followed by medium and high socio-economic status group respondents. Interestingly even in case of wasteful expenditures on items like smoking and drinking etc., the same trend was found to exist. A reverse trend was discernible in case of development expenditure made on farming, dairying and children's education.

Involvement in Farm, Household and Dairying Activities

The operations viz. pesticide dusting, spade work during field irrigation, ploughing, sowing, manure and fertiliser application, uprooting of seedlings and marketing of grains were exclusively performed by men. Transplanting and storage of grains were found to be exclusive domain of women. The other farm operations like weeding, harvesting, carrying head load, threshing and winnowing were performed jointly but predominantly by women.

The average work load of men and women of low socio-economic stratum were higher than those of high and medium socio-economic strata respondents in farm operations. Women of low socio-economic stratum performed more work in crop cultivation than the women of high and medium socio-economic strata. This may be due to the fact that the respondents of low socio-economic stratum are mainly landless and wage earners. They are left with no alternative for their livelihood, therefore, perform more work on other's field as agricultural labourers.

Animal husbandry is predominantly a male affair in case of high socio-economic status (Majority of them had employed permanent male labour to look after the animals), whereas, it is predominantly a female affair in case of farmers of medium and low socio-economic status. On an average, a woman devotes 3.5 hours per day for animal husbandry activities against only 1.6 hours per day devoted by men in these categories.

The household activities are predominantly performed by women irrespective of their socio-economic status and women devote more time in household activities during slack agricultural season. On an average, a woman devotes about 6 hours in peak season against 7 hours in slack season.

Responsibility and Accountability

For the male dominated operations responsibility lies with husbands only and the accountability ranged from 2.89 to 3.28 in high, 2.59 to 3.27 in medium and 2.55 to 3.08 in case of low socio-economic strata respondents. The female dominated operations were the responsibilities of wives in all socio-economic strata. Their accountability ranged from 2.00 to 4.00 in high, 2.62 to 3.62 in medium and 2.40 to 3.52 in case of low socio-economic strata. The jointly performed operations had the joint responsibility of husband and wife in all socio-economic strata. The accountability of such tasks ranged from 2.89 to 3.47 in all socio-economic strata. Thus, the accountability was found to be the highest in case of jointly operated operations followed by female dominated operations. Male dominated operations had the lowest accountability.

Correlates of work Load of Men in cultivation of Major Crops — Wheat and Paddy

Age had very substantial positive direct effect which indicate that aged farmer has a tendency to undertake heavier work load than the younger ones. The education farmers and those coming from joint families have tendency to undertake less work load than those who are less educated and coming from nuclear families. Women of younger age group of lower caste having lower total income, less family education and low land holding were found prone to carry heavier work load of farm work than the women of old age and higher caste who are destined with greater income, more land holding and higher education.

The farmers of higher caste, coming from joint families and having large land holding used to undertake lower load on their farm.

Farm women were so much involved in farming, but they did not have adequate knowledge of modern crop cultivation technology. Whatever, they have learnt about modern technology was by way of their involvement in farming.

Drudgery Undertaken

The various operations/tasks involved in cultivation of wheat and paddy involved varying amounts/magnitudes of drudgery. The total amount of drudgery in carrying out agronomic operations of paddy and wheat cultivation, shouldered by women was found to be significantly higher than that shouldered by men. They seem to be in direct proportion of the time spent in performing various tasks by men as well as women.

The amount of drudgery per man day of work shouldered by

women was found to be significantly higher than that shouldered by men in case of wheat cultivation, but it was in the reverse order in case of rice cultivation. The tasks assigned to men were those which were considered to demand physical power and skill for performance. The tasks assigned to women were those which were considered demanding less physical power, lower skill and more time. They were also monotonous in nature and low paid. The major force influencing inter-sex assignment of tasks was the social value held by men and women alike. The farm machineries in use demonstrated the play of this value. They were all being operated by men and not by women.

Socio-economic correlates of drudgery

Reduction in drudgery with increase in land holding was observed, which may be explained by the two reasons. One was that with the increase in land holding reliance on hired labourer, both on casual and permanent basis, were found to increase who shared part of their drudgery. On the other hand the casual labourers were the landless or marginal/small land holders whose drudgery increased because of their working on the farms of big land holders. Hence, it operated as a double edged weapon. The other reason was that in case of large land holding, different operations carried out by own or hired tractors rather than bullocks which was exclusively done by men. So to that extent drudgery was reduced as against farmers of lower holdings who carried out all such operations by bullock driven ploughs.

Farm women of lower income group used to undertake more drudgery load than the farm women of higher income group.

Contribution of Women in Cultivation of Major Crops

The per acre deployment of labour force reduced with the increase in size of holding consistently in cultivation of all the four major crops. In case of family women labour force, the decrease in man days with increase in size of holding was sharper than that of men labour force in all the four major crops.

Decision-making pattern in the farm family

Men heavily predominated the decision-making processes in farm family in all the six areas related to home management, child rearing, money management, socio-religious rituals, farm and animal management.

Prestige attached to various household tasks

In the eyes of women the tasks which they perform were not so prestigious as the tasks which their husbands performed or they performed with their husbands.

Utilisation pattern of savings

Most of the respondents were able to make some savings from their annual earnings in the immediate preceding year. About half of them had bank deposits of their savings. One-third of them made some assets out of the savings by way of making investment in construction of house, purchase of vehicle, agricultural implements and buffalo etc. A few of them spent on ornaments, social obligations like marriage of children, children's education and health etc.

IMPLICATIONS

Some of the findings reported in the book suggest some serious implications for planners, policy makers, administrators and extension workers. Some of such implications are reported as under:

1. The findings regarding expenditure pattern of the respondents as reported in Table 4.7 make some interesting revelation. The respondents of low income group make a wasteful expenditure of about 4 per cent of their earnings on smoking and drinking (as against only about 1% being spent by the respondents of high income group), whereas they spent meagre one per cent of their earning on education of their children. This calls for a strong need of education to the villagers of low income group in particular to make them aware of this wasteful and health hazardous expenditure and also to inculcate in them the value of education so that they may be able to reverse the pattern of their expenditure.
2. The findings regarding involvement of women in household activities (Table 4.14) revealed that women devoted more time in looking after the children and food preparation etc. during

slack agricultural season but they tend to neglect these tasks during busy agricultural season. This was largely due to economic compulsions and to some extent due to non-availability of agricultural labour. But the consequence of neglecting children and their food etc. is much more serious than the probable consequence on farm. This needs to be brought home through an educational programme and also perhaps mechanisation needs to be introduced to improve the efficiency on farm in peak season in particular so that curtailment of share of time for children is avoided.

3. The findings regarding income as a correlate to work load brought to light the existence of an important social value attached to farm work. Working of family women on farm was found to be valued lower and hence better off families were found to substitute family women labour force with hired labour force. It is only because of the economic compulsions that farmers of lower income allowed their wives to work full time on the farm. This value goes against the dignity which needs to be given to farm work. A suitable educational programme to give dignity to the work they do, needs to be carried out with rural population of Haryana.
4. The findings related to knowledge of farm technology held by farm women revealed that they held about 70 per cent of knowledge in wheat, 56 per cent in paddy and cotton and about 61 per cent in case of *Bajra* cultivation. Although the farm women were found to involve themselves so much in farming, they were not found to have adequate knowledge of modern technology of crop cultivation. This was because there was no deliberate attempt made to give them education about modern technology. Even if some training programmes are organised, it is only the privilege of men to take part in such programmes. This practice needs to be put to an end and involvement of women in farm training programmc have to be made to make agriculture moving.
5. The results regarding drudgery in farm operations do not appear to lend support to the generalisation that drudgery ridden agricultural tasks are assigned to women because of the male dominated social order. The fact that women shoulder greater amount of total drudgery, however, deserves a particular attention. Besides shouldering almost entire work load of

household activities, child rearing and substantial responsibility in dairying, women were found to work in the field for more hours than menfolk. This clearly points to the anormity of work load being carried out by women in the farming community. There is, therefore, a need to make the menfolk in particular, aware of the work load being carried out by women so that they may be more conscious of it and try to strike out a balance.

6. An investigation into the assignment of different cultivation tasks to men and women revealed that prevalence of two social values played a significant role in it. They are: (a) women are physically less powerful than men and (b) they have lower ability to acquire skill. Men may be physically stronger than women in absolute term but what was ignored was the examination whether the amounts of power the particular tasks required were also possessed by women or not. This suggests a very serious implication to technological solutions if innovated to emancipate women from the drudgery they undertake in farm operations/tasks. Tools and techniques may be devised to raise efficiency and reduced time and energy consumption in performance of the task assigned to women. But then they may ungrudgingly hand over the task to men believing that they require higher level of skill.

   It is, therefore, absolutely essential that men and women are given education to help them to explode the myth that women are *abala* (powerless) and have lower ability to acquire skills.

7. The study has made it amply clear that farm related activities are performed almost equally by the womenfolk (as by the menfolk), but unfortunately their contribution in decisions related to farming was found to be very meagre. Somehow the cultural value that 'Farm work is the domain of men an household of women' seems to prevail in the society without any regard paid to the contribution made by women in the family. This value seems to influence decisions about farm related activities taken in farmers' family, but this does not seem to influence the decisions regarding home management activities, where men were found to poke their nose and influence about half of the decisions (about 27% independently and 21% jointly with women). Even in case of other areas of decision-making domination of men is quite discernible. Yet, there was no sign of dissent from womenfolk against this

domination. Male chauvinism obviously smacks from this. Therefore, there is a need to devise an educational programme for both men and women in the society to enhance respectability for women in farming and the society without disturbing the prevailing harmony.

8. An investigation into the prestige attached to the tasks performed by husbands and wives clearly pointed out that in the eyes of women the tasks which their husbands perform along with them or independently were significantly more prestigious. This speaks of the sense of inferiority, the farm womenfolk suffers from in our society. There is a need to pay attention to this so that the sense of inferiority goes from them. Without empowering them the dream of social growth and development cannot be realised.

APPENDIX I

Socio-Economic Status Scale (Rural)

Name of the respondent ____________________ w/o/s/o ________________

Age _________ years. Head of the family ________________________

Village ______________________ District ______________________

1. *Caste*	Detailed scoring
Schedule caste	1
Lower caste	2
Artisan caste	3
Agricultural caste	4
Prestige caste	5
Dominant caste	6
2. *Occupation*	
Agricultural labour	1
Caste occupation	2
Independent profession/business	3
Cultivation	4
Service	5

3. *Education*		Male Head	Female Head	Children 1 2 3 4 5 6 7	Other members
Illiterate	0				
Primary	1				
Middle	2				
High School	3				
Technical/vocational education	4				
Graduate	5				
Post-graduate and above	6				

Contd.

Contd.

	Male Head	Female Head
4. *Social Participation*		
No participation	0	0
Member of one organisation	1	1
Member of more than one organisation	2	2
Office holder	3	3
Wider Public	6	6
5. *Size of Family*		
Up to 5 members	1	
Above 5 members	2	
6. *Type of family*		
Nuclear	1	
Joint	2	
7. *Land*		
No Land	0	
Up to 2.5 acre	1	
2.5 - 5.0 acre	2	
5.0 - 7.5 acre	3	
7.5 - 15.0 acre	4	
Above 15 acre	5	
8. *House*		
Katcha	1	
Mixed	2	
Pucca	3	
9. *Farm Power*		
1 - 2 drought animals	1	
3 - 4 drought animals	2	
5 - 6 and above	3	
Tractor	6	
10. *Improved Agricultural Implements*		
Up to 4	4	
4 and above	5	
Tractor driven implements	6	
11. *Milch Animals*		
1 - 2 milch animals	1	
3 - 4 milch animals	2	
5 - 8 milch animals	3	
9 and above	4	

Contd.

12. *Material Possession*	
(a) *Household Appliances*	
Stove	1
Coal iron	1
Sewing machine	1
Electric iron	1
Cycle	1
Radio	1
(b) *Prestige items*	
Electric Iron	2
Electric heater	2
Mixer blender/grinder	2
Food processor	2
Juicer	2
Toaster	2
Oven	2
Immerson rod	2
Pressure cooker	2
Improved storage bins	2
Gobar gas plant	2
Gas *chullah*	2
Solar cooker	2
Sofa set	2
Dinning set	2
Double bed	2
Knitting machine	2
Tape recorder/record player	2
Twin player-cum-recorder	2
Scooter/Motor cycle	2
Refrigerator	3
Cooler	3
Washing machine	3
Television	3
V.C.R.	3
Camera	3
Geyser	3
Car/Jeep/Matador	3
Any other	3

APPENDIX II

Socio-economic status of selected villages of Karnal and Hisar districts.

1 Gogripur Village (Karnal District)

Sl. No.	Socio-economic status score	Frequency	Weighted frequency	Weighted cumulative frequency
1	2	3	4	5
1.	4	5	20	20
2.	5	14	70	90
3.	6	23	138	228
4.	7	28	196	424
5.	8	17	136	569
6.	9	12	108	668
7.	10	9	90	758
8.	11	19	209	967
9.	12	22	264	1231
10.	13	7	91	1322
11.	14	16	224	1546
12.	15	7	105	1651
13.	16	8	128	1779
14.	17	9	163	1942
15.	18	5	90	2032
16.	19	6	114	2146
17.	20	5	100	2246
18.	21	9	189	2435
19.	22	10	220	2655
20.	23	17	391	3055
21.	24	20	480	3526
22.	25	15	375	3900
23.	26	25	650	4550
24.	27	7	189	4739
25.	28	18	504	5243
26.	29	10	290	5533
27.	30	15	450	5983
28.	31	12	372	6355
29.	32	8	256	6611
30.	33	10	330	6941
31.	34	14	476	7417
32.	35	11	385	7802
33.	36	8	288	8090

Contd.

Contd.

1 2	3	4	5	
	1. Gogripur village (Karnal district)			
34.	37	12	444	8534
35.	38	16	608	9142
36.	39	14	546	9688
37.	40	5	200	9888
38.	41	4	164	10052
39.	42	6	252	10304
40.	43	4	172	10476
41.	44	1	44	10520
42.	45	4	180	10700
43.	46	4	184	10884
44.	47	2	94	10978
45.	48	1	48	11026
46.	49	3	147	11173
47.	50	2	100	11273
48.	51	1	51	11324
49.	52	0	0	11324
50.	53	1	53	11377
51.	54	1	54	11431
52.	57	2	114	11545
53.	61	1	61	11606
54.	62	1	62	11668
55.	65	1	65	11733
56.	80	1	80	11813
57.	89	1	89	11902
	2. Mehmoodpur Village (Karnal district)			
1.	5	4	20	20
2.	6	8	48	68
3.	7	7	49	117
4.	8	9	72	189
5.	9	6	54	243
6.	10	11	110	353
7.	11	13	143	496
8.	12	8	86	592
9.	13	9	117	709
10.	14	7	98	807
11.	15	16	240	247
12.	16	12	192	1239
13.	17	6	102	1341
14.	18	14	252	1593

Contd.

Contd.

1 2	3	4	5	
15.	19	9	171	1764
16.	20	10	200	1964
17.	21	20	420	2384
18.	22	15	320	2704
19.	23	11	253	2957
20.	24	12	288	3245
21.	25	8	200	3445
22.	27	10	270	3715
23.	28	11	308	4023
24.	30	6	180	4203
25.	31	15	465	4668
26.	32	13	416	5084
27.	33	9	297	5381
28.	34	6	204	5585
29.	35	8	280	5865
30.	36	10	360	6225
31.	37	13	481	6706
32.	38	6	228	6934
33.	39	7	273	7207
34.	40	5	200	7407
35.	41	3	123	7530
36.	42	4	168	7698
37.	43	2	86	7784
38.	44	5	220	8004
39.	45	1	45	8049
40.	46	1	46	8095
		3. Shahpur Village (Hisar district)		
1.	4	3	12	12
2.	5	18	90	102
3.	6	42	252	354
4.	7	25	150	504
5.	8	30	240	744
6.	9	17	153	897
7.	10	14	140	1037
8.	11	12	132	1169

Contd.

Contd.

1	2	3	4	5
9.	12	19	228	3397
10.	13	19	247	1644
11.	14	19	266	1910
12.	15	13	195	2105
13.	16	21	336	2441
14.	17	16	272	2713
15.	18	14	252	2965
16.	19	16	285	3250
17.	20	11	220	3470
18.	21	7	147	3617
19.	22	6	132	3739
20.	23	8	184	3933
21.	24	8	192	4125
22.	25	4	100	4225
23.	26	9	234	4459
24.	27	2	254	4513
25.	28	3	84	4597
26.	29	1	29	4626
27.	30	3	90	4716
28.	31	5	155	4871
29.	32	3	96	4967
30.	33	1	33	5000
31.	34	1	34	5034
32.	35	0	0	5034
33.	36	0	0	5034
34.	37	0	0	5034
35.	38	0	0	5034
36.	39	3	117	5151
37.	40	1	40	5191
38.	41	1	41	5232
39.	42	0	0	5232
40.	43	0	0	5232
41.	44	1	44	5276
42.	45	0	0	5276
43.	46	0	0	5276
44.	47	0	0	5276
45.	48	1	48	5324
46.	49	0	0	5324
47.	50	0	0	5324
48.	51	0	0	5324
49.	52	0	0	5324
50.	53	0	0	5324

Contd.

Contd.

1	2	3	4	5
51.	54	0	0	5324
52.	55	0	0	5324
53.	56	1	56	5380
54.	57	0	0	5380
55.	58	0	0	5380
56.	59	1	59	5439
		4. Patna Village (Hisar district)		
1.	8	1	8	8
2.	9	0	0	8
3.	10	6	60	68
4.	11	5	55	123
5.	12	9	108	231
6.	13	9	117	348
7.	14	12	168	516
8.	15	13	195	711
9.	16	8	128	839
10.	17	10	117	1009
11.	18	7	126	1135
12.	19	12	228	1363
13.	20	6	120	1483
14.	21	12	252	1735
15.	22	7	154	1889
16.	23	4	92	1981
17.	24	6	144	2125
18.	25	8	200	2325
19.	26	9	234	2555
20.	27	11	297	2856
21.	28	6	168	3024
22.	29	5	145	2169
23.	30	5	150	3319
24.	31	8	248	3567
25.	32	6	192	3759
26.	33	4	132	3891
27.	34	6	204	4095
28.	35	7	245	4340
29.	36	3	108	4448
30.	37	6	222	4670
31.	38	6	228	4898
32.	39	8	31	5210

Contd.

Contd.

1	2	3	4	5
33.	40	6	240	5450
34.	41	4	164	5614
35.	42	4	178	5792
36.	43	10	430	6222
37.	44	3	132	6354
38.	45	4	180	6534
39.	46	3	138	6672
40.	47	2	94	6766
41.	48	1	48	6814
42.	49	1	49	6863
43.	50	0	0	6863
44.	51	2	102	6965
45.	52	0	0	6965
46.	53	0	0	6965
47.	54	2	108	7073
48.	55	0	0	7073
49.	56	1	56	7129
50.	57	0	0	7129
51.	58	1	58	7187
52.	59	0	0	7187
53.	60	1	60	7247
54.	61	0	0	7247
55.	62	1	62	7309
56.	63	0	0	7309
57.	64	0	0	7309
58.	65	1	65	7374
59.	66	1	66	7440

APPENDIX III

Interview Schedule

1. Details of operations carried out by men and women in crop cultivation (wheat, paddy, cotton and Bajra).

Sl. No.	Farm Operations	Total Labour days/acre		Type of equipments Implements	Family Labour/ acre		Hired labour/ acre		Total hours/day	
		Male	Female		Male	Female	Male	Female	Peak season	Stack season
1	2	3	4	5	6	7	8	9	10	11

1. Land Preparation
2. Raising nursery
3. Direct sowing
4. Transplanting
5. Manure and Fertiliser applications
6. *Inter-cultural operations*

   (a) Weeding by *Khurpi*
   (b) Weeding by *Kasola*
   (c) Thinning
   (d) Gap-filling

7. Irrigation
8. Plant Protection measures
9. Harvesting

   (a) Cutting
   (b) Picking

10. Shifting produce to Threshing floor
11. Threshing
12. Winnowing
13. Keeping part of grains for consumption
14. Sun drying
15. Marketing of grains

2. Details of Operations Carried Out by Men and Women for Animal Husbandry and Dairying

Sl. No.	Operations	Total No. of animals	Family Labour/ day		Hired Labour/ day		Hours spent/ day
			Male	Female	Male	Female	
1	2	3	4	5	6	7	8
1.	Bringing fodder from field						
2.	Chaffing the fodder						
3.	Prepare feed for animals						
4.	Grazing animals						
5.	Health care of animals (Dehorning and deworming etc.)						
6.	Offering water to animals						
7.	Cleaning of animals						
8.	Cleaning of shed and preparing cow-dung cakes						
9.	Milking						
10.	Selling of milk						
11.	Cleaning of utensils used for milking						
12.	*Preparing milk products*						
	(a) Ghee						
	(b) Paneer						
	(c) Any other						
13.	Any other (specify)						

3. Details of Operations Carried out by Men and Women in Household Activities.

Sl. No.	Operations	Male	Female	Family Labour		Hired Labour				Total hours per day		Income
				M	F	Peak season		Slack season		Peak season	Slack season	
						M	F	M	F			
1	2	3	4	5	6	7	8	9	10	11	12	13

1. Cooking
2. Fetching water
3. Collecting fuel wood and twigs etc.
4. Collection of cow-dung and preparation of cow-dung cakes/ coal dust balls
5. Child care
6. *Routine household work*
   - (a) Cleaning the house and utensils
   - (b) Washing clothes and house
7. *Food preparation*
   - (a) Papad making
   - (b) Pickle making
   - (c) Squash, jam, jelly making
   - (d) Any other
8. *Grain processing*
   - (a) Parboiling of rice
   - (b) Dehusking
   - (c) Grinding

Contd.

Contd.

1	2	3	4	5	6	7	8	9	10	11	12	13
9.	Grain storage											
10.	*Income generating activities*											
	(a) Chalk making											
	(b) Candle making											
	(c) Durri weaving											
	(d) Pickle making											
	(e) Papad and wadi making											
	(f) Hand knitting											
	(g) Machine knitting											
	(h) Embroidery											
	(i) Crochet work											
	(j) Stitching of clothes											
	(k) Mat weaving											
11.	Any other											

4. (a) During peak season, women do work for long hours in agriculture then how they manage their household activities?
   - By over working
   - By reducing household activity
   - Any other

   (b) During slack season, women do work for less hours in agriculture then how they utilise their time.
   - Spend more time in household activities
   - Use leisure time for income generating activities
   - Any other

5. **Responsibility and accountability of men and women for various farm operations**

Sl. No.	Operations	Responsibility				Accountability				
		Hus-band	Wife	Joint-	Any other	5	4	3	2	1
1	2	3	4	5	6	7	8	9	10	11
1.	Land preparation									
2.	Nursery raising									
3.	Direct sowing									
4.	Transplanting									

5. Manure and fertiliser application
6. *Inter cultural operations*

   (a) weeding
   (b) thinning
   (c) gap-filling

7. Irrigation
8. Plant Protection measures
9. *Harvesting*

   (a) cutting
   (b) picking

10. Shifting produce to threshing floor
11. Threshing
12. Winnowing
13. Keeping part of grains for consumption
14. Sundrying
15. Marketing of grains

6. Knowledge Inventory of Farm Women for Improved Crop Technology

Wheat Cultivation

(i) Do you know the name of high yielding varieties of wheat?
Timely sowing varieties————————
Late sowing varieties————

(ii) Do you know the name of chemical for seed treatment?
Name————quantity for one kg seed————gm.

(iii) Do you know the optimum time of sowing?

(iv) Seed rate
Timely sowing crops————
Late sowing crops————

(v) Do you know the spacing for timely sown crops?

(vi) Do you know the depth of sowing?

(vii) Do you know the distance between plants?

(viii) Do you know the quantity of fertiliser/acre?

(a) Nitrogenous fertiliser————
(b) Phosphatic fertiliser————
(c) Potash fertiliser————

(ix) After how many days of sowing the first dose of nitrogen should be applied to crop————days.

(x) Do you know the quantity of urea for spraying on one acre crop————kg.

(xi) How many days after sowing the following irrigation should be given?

(a) First irrigation————days
(b) Second irrigation————days
(c) Third irrigation————days
(d) Fourth irrigation————days

(xii) Do you know the method of irrigation?

(xiii) Do you know the name of weedicide to control weeds?

Name of the weedicide	*State of plant*	*Dose per acre*	*Weeds to be controlled*

(xiv) Name of fungicide to control rust

(a) Name————
(b) Quantity for one acre————gms.
(c) Days interval to repeat application————days.

Paddy Cultivation

(i) Do you know the names of high yielding varieties of paddy?
Timely sowing varieties————
Late sowing varieties————

(ii) What is the seed rate of improved paddy varieties?
Timely sowing varieties————
Late sowing varieties————

(iii) When is the paddy nursery raised?

(iv) What is the fertiliser requirement of the nursery area ment for one acre of transplantation?
Compost————
Nitrogenous fertiliser————
Phosphatic fertiliser————

(v) Do you know the advantages of chemical seed treatment? Yes/No
If yes, what are the advantages of seed treatment?

(vi) Do you know the name of the chemical used for seed treatment?

(vii) Do you know the dose of chemical required for treating seed for one acre?

(viii) How many days old seedlings are used for transplantation?

(ix) Recommended plant spacing————

(x) Depth of planting————

(xi) Number of seedlings per hill————

(xii) Recommended period for gap-filling————

(xiii) Quantity of Basal dose fertiliser per acre
Nitrogenous fertiliser————
Potassic fertiliser————

(xiv) Method of spraying urea————

(xv) How many days after the transplantation nitrogenous fertiliser should be applied to the crop?

	Particulars	*Number of days after transplanting*
1.	1st dose of nitrogen	
2.	2nd dose of nitrogen	
3.	3rd dose of nitrogen	
4.	4th dose of nitrogen	

(xvi) Do you know the names of insects/diseases and recommended chemicals and dose to control them in one acre field.

	Name of insects/diseases	*Name of chemicals*	*Dose of chemical for one acre field*
1.			
2.			
3.			

(xvii) Do you know the herbicide, dose and stage of crop growth at the time of use?

Name of herbicide	*Amount per acre*	*Stage of crop*	*Growth*

Cotton Cultivation

(i) Do you know the names of high yielding varieties of cotton?
Timely sowing varieties————
Late sowing varieties————

(ii) What is the seed rate of cotton crop?
Timely sowing varieties————
Late sowing varieties————

(iii) Do you know the name of chemical used for seed treatment?
Name of chemical————
Quantity for one kg seed————gm.

(iv) Do you know the optimum time of sowing?
Timely sowing varieties————
Late sowing varieties————

(v) Do you know the depth of seed for sowing?
(vi) Do you know the distance between plants?
(vii) Do you know the quantity of fertiliser/acre?
Nitrogenous fertiliser—————
Phosphatic fertiliser—————
Potash fertiliser—————
(viii) After how many days of sowing the first dose of nitrogen should be applied to crop—————days
(ix) Do you know the quantity of urea for spraying on one acre crop—————kg.
(x) Method of fertiliser application.
(xi) How many days after sowing the following irrigation should be given?
(a) Ist irrigation—————days
(b) 2nd irrigation—————days
(c) 3rd irrigation—————days.
(xii) Method of irrigation.
(xiii) Do you know the weedicide to control weeds?

Name of the weedicide	*State of plant*	*Dose per acre*	*Weeds to be controlled*

(xiv) Name of fungicide to control rust
(a) Name—————
(b) Quantity for one acre—————gms.
(c) Days interval to repeat application—————days.
(xv) Do you know the name of insects and pests and their treatment?
(a) Name of insects/pests—————
(b) Treatment—————

Bajra Cultivation

(i) Do you know the names of High yielding varieties of Bajra?
Timely sowing varieties—————
Late sowing varieties—————
(ii) Do you know the seed rate?
Timely sowing varieties—————
Late sowing varieties—————
(iii) Do you know the name of chemical for seed treatment?
Name of chemical—————
Quantity for one kg seed—————gm.
(iv) Do you know the advantages of seed treatment?
(v) Do you know the optimum time of sowing?
Timely sowing varieties—————
Late sowing varieties—————
(vi) What is the depth of seed for sowing?

(vii) Do you know the distance between plants?

(viii) Do you know the requirement of fertiliser per acre?
Nitrogenous fertiliser————
Phosphatic fertiliser————
Potash fertiliser————

(ix) After how many days of sowing the first dose of nitrogen should be applied to crop————days.

(x) Do you know the quantity of urea for spraying on one acre crop————kg.

(xi) What is the method of applying fertiliser?

(xii) How many days after sowing irrigation should be given?
(a) 1st irrigation————days
(b) 2nd irrigation————days
(c) 3rd irrigation————days

(xiii) Do you know different method of irrigation?

(xiv) Do you know the name of weedicide and application of weedicide?

Name of weedicide	*Stage of plant*	*Dose per acre*	*Weeds to be controlled*

(xv) Name the fungicide to control rust

(a) Name————
(b) Quantity for one acre————gms.
(c) Days interval to repeat application————days.

(xvi) Do you know the name of insects, pests and their treatment?
Name of the Insect/pests————Treatment————

7. Impact of Mechanisation on Women

(a) Do you use the following implements on your field or in your village?

Sl. No.	*Implements*	*Own field*	*In village*
1.	Tractor		
2.	Thresher		
3.	Combine harvester		
4.	Seed-cum fertiliser drill		
5.	Transplanter		

1. What are the operations being done by tractor?
   - ploughing
   - threshing

- carrying load
- sowing
- others

(a) Which are the above mentioned operations done by women and now they are not getting opportunity due to tractor?

Crop	*Operations*	*Women labourer displaced per acre*
Kharif		
Rabi		
Total		

(b) In your opinion how many————women labourers displaced per acre in a year.

(c) What is the effect of displacement on women?
- Reduced income
- Made idle
- Relieve from drudgery
- Any other

(d) After displacement, how are they using their spared time and energy?
- Take rest
- Found alternative earning engagement
- Devote more time to household activities

2. What are the operations being done by thresher?
   - dehusking

(a) Which are the above mentioned operations done by women and they are not getting opportunity due to thresher?

Crop	*Operations*	*Women labourer displaced per acre*
Kharif		
Rabi		
Total		

(b) In your opinion how many————women labourers displaced per acre in a year.

(c) What is the effect of displacement on women?
- Reduced income
- Made idle
- Relieve from drudgery
- Any other

(d) After displacement, how are they using their spared time and energy?
- Take rest
- Found alternative earning engagement
- Devote more time to household activities

3. What are the operations being done by combine harvester?
- Harvesting
- dehusking

(a) Which are the above mentioned operations done by women and they are not getting opportunity due to combine harvester?

Crop	*Operations*	*Women labourer displaced per acre*
Kharif		
Rabi		
Total		

(b) In your opinion how many————women labourers displaced per acre in a year.

(c) What is the effect of displacement on women?
- Reduced income
- Made idle
- Relieve from drudgery
- Any other

(d) After displacement, how are they using their spared time and energy?
- Take rest
- Found alternative earning engagement
- Devote more time to household activities

4. What are the operations being done by seed cum fertilizer drill?
- Sowing
- Fertilizer application
- Sowing in lines

(a) Which are the above mentioned operations done by women and they are not getting opportunity due to seed cum fertilizer drill?

Crop	*Operations*	*Women labourer displaced per acre*
Kharif		
Rabi		
Total		

(b) In your opinion how many————women labourers displaced per acre in a year.

(c) What is the effect of displacement on women?
- Reduced income
- Made idle
- Relieve from drudgery
- Any other

(d) After displacement, how are they using their spared time and energy?
- Take rest
- Found alternative earning engagement
- Devote more time to household activities

5. What are the operations being done by transplanter?
- Transplanting of seedlings

(a) Which are the above mentioned operations done by women and they are not getting opportunity due to transplanter?

Crop	*Operations*	*Women labourer displaced per acre*
Kharif		
Rabi		
Total		

(b) In your opinion how many————women labourers displaced per acre in a year.

(c) What is the effect of displacement on women?
- Reduced income
- Made idle
- Relieve from drudgery
- Any other

(d) After displacement, how are they using their spared time and energy?
- Take rest
- Found alternative earning engagement
- Devote more time to household activities.

8. Inventory for Farm and Home Related Decisions

Sl. No.	*Decisions*	*Only Husband*	*Only Wife*	*Husband and wife jointly*
1	2	3	4	5

1. *Home Management Related Decisions*
   (i) Construction of a new house
   (ii) Construction of movable parts like racks, chakki, chullah, hara.
   (iii) Getting the house white washed.
   (iv) Decorating the house on festivals.

(v) Preparing fancy articles like dolls, basket, embroidery etc.

2. *Child Related Decisions*
   (i) Settling children's arguments and fights
   (ii) Deciding for children's purchases.
   (iii) Deciding about school to which children should go
   (iv) Age at which children should go to school
   (v) Level of education.
   (vi) Type of education general or technical
   (vii) Deciding about occupation for son/daughter
   (viii) Marriage age of children
   (ix) Selecting a match for son/daughter
   (x) Amount of dowry to be given
3. *Money Related Decisions*
   (i) How much money should spend on different items?
   (ii) Buying and selling of different articles
   (iii) Buying farm inputs
   (iv) Buying farm machinery
   (v) Buying daily necessities like clothes, luxury items etc.
   (vi) Buying costly items like fan cooler, TV, refrigerator, scooter etc.
   (vii) Selling of farm produce
   (viii) Selling of home produce
   (ix) Taking and giving loans
   (x) Deciding manner of savings
   (xi) Insurance of family members
   (xii) Insurance of crops
   (xiii) Insurance of animals
   (xiv) Buying and selling of animals
   (xv) Selling of extra lands
4. *Socio-Religious Rituals Related Decisions*
   (i) When to call and send back daughters to their in-laws place
   (ii) Attending marriage, births and death ceremonies
   (iii) Going to school gatherings
   (iv) Going to Mahila Mandal
   (v) Kind and amount of charity to be given
   (vi) When and whom to worship
   (vii) Going to religious gatherings — Bhajan Mandali
   (viii) Giving donation
5. *Farm Related Decisions*
   (i) How many crops and area under each crop?
   (ii) What crops and their varieties to be grown?
   (iii) Seed related decisions
   (iv) Fertilizer related decisions

(v) How many times the field should be irrigated?
(vi) How many channels to be constructed
(vii) Install tube-wells, pump or 'rahat'
(viii) Plant protection related decisions

6. *Animal Related Decisions*

(i) How many cattles to be kept?

9. Given Below is a List of Tasks that have to be Done, will you Please Tell Who Perform the Following Family and Household Tasks and Prestige Attached with These Tasks

Sl. No.	*Tasks*	*Who perform tasks*			*Prestige attached to the tasks*				
		Only Wife	*Only Husband*	*Wife and husband together*	*5*	*4*	*3*	*2*	*1*
1	2	3	4	5	6	7	8	9	10
1.	*Food Related Tasks*								
(i)	Churning								
(ii)	Grinding								
(iii)	Fetching water								
(iv)	Collecting fuel and set to fire								
(v)	Daily cooking								
(vi)	Cooking on festivals								
(vii)	Serving food								
(viii)	Distribution of milk and milk products								
(ix)	Carrying meal to the field								
(x)	Cleaning grains								
(xi)	Cleaning utensils								
(xii)	Storing grains								
(xiii)	Food preservation								
(xiv)	Use of high yielding variety for consumption								
2.	*Household Tasks*								
(i)	Brooming and Mopping								
(ii)	Mending clothes								
(iii)	Washing clothes								
(iv)	Stitching clothes								
(v)	Knitting and Embroidery								
(vi)	Spinning								
(vii)	Rearranging household activities								
(viii)	Bed making								
(ix)	Daily shopping								
3.	*Child Related Tasks*								
(i)	Bathing children								
(ii)	Dressing children								
(iii)	Feeding children								
(iv)	Playing or talking with children								
(v)	Teaching children household tasks								
(vi)	Teaching children good habits								

(vii) Treatment of sick children
(viii) Taking children to doctor
(ix) Taking and sending children

4. *Money Management Related Tasks*
(i) Paying loans
(ii) Keeping records
(iii) Paying bills
(iv) Keeping money

5. *Socio Religious Tasks*
(i) Taking care of guests
(ii) Helping neighbours and friends
(iii) Taking initiative and performing puja on different occasions
(iv) Keeping fast on religious occasions

10. Different Sources of Income

Sl. No.	*Source*	*Income in Rupees*
1.	Selling of milk and milk products	
2.	Field crops	
3.	Agricultural labourers	
4.	Service	
5.	Any other	

11. Expenditure on Different Field and Household Items

Different Expenditure	*Expenditure in rupees*
1. *Expenditure on Maintenance*	
(i) Food	
(ii) Fuel	
(iii) Clothes for females	
(iv) Clothes for other members of family	
(v) Renovation or rent of house	
(vi) Medicines	
2. *Expenditure on development*	
(i) Education	
(ii) Seed, fertilizer, insecticides for field crops	
(iii) Irrigation	
(iv) Labour in cash to labourers	
(v) Cattle feed	
3. *Expenditure on Wasteful Items*	
(i) Smoking and liquor etc.	
(ii) Miscellaneous	
(iii) Any other	

12. **How Did You Utilize the Savings of Preceding Year?**

- Marriage
- Purchase of buffaloes
- Agricultural implement (Specify)
- Ornaments
- House construction
- Bank deposits
- Education of son or daughters
- Children's health
- Clothes for women
- Good diet for women
- Raises socio-economic status
- Good treatment of women

13. How Do You Utilize the Earnings Made by Women Through Income Generating Projects?

- Giving to husbands
- Keeping with themselves
- Expenditure on children's care
- Expenditure on children's education
- Any other (specify)

APPENDIX-IV

A Semi Projective Instrument Used for Drudgery Index

Plate 1: Pesticide Dusting

Plate 2: Harvesting

Plate 3: *Carrying Load on Head*

Plate 4: Spade Work During Field Irrigation

Plate 5: Transplanting

Plate 6: Weeding by 'Kasola'

Plate 7: Weeding by 'Khurpi'

Plate 8: Sowing behind Plough

Plate 9: Bullock Ploughing

Plate 10: Puddling by Plough

APPENDIX-V

Fij Matrix of Ladies Drudgery

	1	2	3	4	5	6	7	8	9	10
1.	—	22	17	19	27	28	27	26	24	23
2.	8	—	19	28	28	25	23	25	23	27
3.	13	11	—	24	26	28	24	25	22	20
4.	11	2	6	—	21	28	29	17	19	23
5.	3	2	4	9	—	26	20	20	23	24
6.	2	5	2	2	4	—	26	28	22	27
7.	3	7	6	1	10	4	—	21	19	26
8.	4	5	5	13	10	2	9	—	24	18
9.	6	7	8	11	7	8	11	6	—	20
10.	7	3	10	7	6	3	4	12	10	—

P_{ij} Matrix of Ladies Drudgery

	1	2	3	4	5	6	7	8	9	10
1.	—	.73	.56	.63	.90	.93	.90	.86	.80	.76
2.	.27	—	.63	.93	.93	.83	.76	.83	.76	.90
3.	.44	.37	—	.80	.86	.93	.80	.83	.73	.66
4.	.37	.07	.20	—	.70	.93	.96	.56	.63	.76
5.	.10	.07	.14	.30	—	.86	.66	.66	.76	.80
6.	.07	.17	.07	.07	.14	—	.86	.93	.73	.90
7.	.10	.24	.20	.04	.34	.14	—	.70	.63	.86
8.	.14	.17	.17	.44	.34	.07	.30	—	.80	.60
9.	.20	.24	.27	.37	.24	.27	.37	.20	—	.66
10.	.24	.10	.34	.24	.20	.10	.14	.40	.34	—

Zij Matrix of Ladies Drudgery
(Case V Model)

	1	2	3	4	5	6	7	8	9	10
1.	—	.613	.151	.332	1.282	1.476	1.282	1.080	.842	.706
2.	-.613	—	.332	1.476	1.476	.954	.706	.954	.706	1.282
3.	-.151	-.332	—	.842	1.080	1.476	.842	.994	.613	.412
4.	-.332	-1.476	-.842	—	.524	1.476	1.751	.151	.332	.706
5.	-1.282	-1.476	-1.080	-.524	—	1.080	.412	.412	.706	.824
6.	-1.476	-.954	-1.476	-1.476	-1.080	—	1.080	1.476	.513	1.282
7.	-1.282	-.706	-.842	1.751	-.412	-1.080	—	.524	.332	1.080
8.	-1.080	-.954	-.994	-.151	-.412	-1.476	-.524	—	.842	.253
9.	-.842	-.706	-.613	-.332	-.706	-.613	-.332	-.842	—	.412
10.	-.706	-1.282	-.412	-.706	-.824	-1.282	-1.080	-.253	-.412	—
ΣZ_{ij}	-7.764	-7.273	-5.776	-2.290	+0.928	+2.011	+4.137	+4.496	+4.574	+6.957
$\Sigma Z_{ij}/n$	-.7764	-.7273	-.5776	-.2290	+.0928	+.2011	+.4137	+.4496	+.4574	+.6957
Add highest negative value to each cell	-.7764 +	-.7764 +	-.7764 +	-.7764 +	-.7764 +	-.7764 +	-.7764 +	-.7764 +	-.7764 +	-.7764 +
Scale Values	0	+.0491	+.1988	+.5474	+.8692	+.9775	+1.1901	+1.2260	+1.2338	+1.4721

New Z Matrix of Ladies Drudgery
(Case III Model)

	1	2	3	4	5	6	7	8	9	10
1.	—	.961	.237	.520	2.313	2.009	2.009	1.692	1.319	1.106
2.	.582	—	.315	1.402	.906	1.402	.671	.906	.671	1.218
3.	1.184	-.405	—	1.026	1.799	1.316	1.026	1.212	.747	.502
4.	-.109	-.484	-.276	—	.484	.172	.574	.049	.109	.231
5.	-0.087	-.056	-.087	.087	—	.064	.163	.163	.279	.325
6.	-0.506	-.583	-.427	-.207	+.427	—	.064	.087	.036	.076
7.	-0.501	-.276	-.329	-.685	-.422	.161	—	.205	.130	.422
8.	-0.916	-.809	-.843	-.182	-1.252	.349	-.444	—	.714	.214
9.	-1.879	-1.576	-1.368	-.741	-1.368	-1.575	-.741	-1.879	—	.919
10.	-1.415	-2.570	-.826	-.1.415	-2.570	-1.752	-2.165	-.507	.826	—
ΣZ_{ij}	-7.179	-5.798	-3.604	-.315	+.317	+1.097	+1.157	+1.192	+3.179	+5.013
$\Sigma Z_{ij}/n$	-.7179	-.5798	-.3604	-.0315	+.0317	+.1097	+.1157	+.1192	+.3179	+.5013
Add highest negative value to each cell	-.7179 +	-.7179 +	-.7179 +	-.7179 +	-.7179 +	-.7179 +	-.7179 +	-.7179 +	-.7179 +	-.7179 +
Scale values	0	+.1381	+.3575	+.6864	+.7496	+.8276	+.8336	+.8371	+1.0358	+1.2192

F_{ij} Matrix of Gents Drudgery

	1	2	3	4	5	6	7	8	9	10
1.	—	21	24	28	27	22	26	17	22	22
2.	9	—	22	17	20	17	25	23	22	20
3.	6	8		22	21	21	24	25	20	20
4.	2	13	8	—	17	22	21	23	21	28
5.	3	10	9	13	—	28	23	17	20	19
6.	8	13	9	8	2	—	24	21	19	24
7.	4	5	6	9	7	6	—	23	27	16
8.	13	7	5	7	13	9	7	—	20	17
9.	8	8	10	9	10	11	3	10	—	18
10.	8	10	10	2	11	6	14	13	12	—

P_{ij} Matrix of Gents Drudgery

	1	2	3	4	5	6	7	8	9	10
1.	—	.70	.80	.93	.90	.73	.86	.57	.73	.73
2.	.30	—	.73	.57	.66	.57	.83	.76	.73	.66
3.	.20	.27	—	.73	.70	.70	.80	.83	.66	.66
4.	.07	.43	.27	—	.57	.73	.70	.76	.70	.93
5.	.10	.34	.30	.43	—	.93	.76	.57	.66	.63
6.	27	.43	.20	27	.07	—	.80	.70	.63	.80
7.	.14	.17	.20	.30	.24	.20	—	.76	.90	.53
8.	.43	.24	.17	.24	.43	.30	.24	—	.66	.56
9.	.27	.27	.34	.30	.34	.37	.10	.34	—	.60
10.	.27	.34	.34	.07	.37	.20	.47	.44	.40	

Zij Matrix of Gents Drudgery
(Case V Model)

	1	2	3	4	5	6	7	8	9	10
1.	—	.524	.842	1.476	1.282	.613	1.080	.176	.613	.613
2.	-.524	—	.613	.176	.412	.176	.954	.706	.613	.412
3.	-.842	-.613	—	.613	.524	.524	.842	.954	412	.412
4.	-1.476	-.176	-.613	—	.176	.613	.524	.706	.524	1.476
5.	-1.282	-.412	-.524	-.176	—	1.476	.706	.176	.412	.332
6.	-.513	-.176	-.524	-.613	-1.476	—	.842	.524	.332	.842
7.	-1.080	-.954	-.842	-.524	-.706	-.842	—	.706	1.282	.075
8.	-.176	-.706	-.954	-.706	-.176	-.524	-.706	—	.412	.151
9.	-.613	-.613	-.412	-.524	-.412	-.332	-1.282	-.412	—	.253
10.	-.613	-.412	-.412	-1.476	-.332	-.842	-.075	-.151	-.253	—
ΣZ_{ij}	-7.219	-3.538	-2.826	-.754	-.708	-.138	+2.885	+3.385	+4.347	+4.566
$\Sigma Z_{ij}/n$	-.7219	-.3538	-.2826	-.0754	-.0708	-.0138	+.2885	+.3385	+.4347	+.4566
Add highest negative value	.7219	-.7219	-.7219	-.7219	-.7219	-.7219	-.7219	-.7219	-.7219	-.7219
to each cell	+	+	+	+	+	+	+	+	+	+
Scale value	0	+.3681	+.4393	+.6465	+.6511	+.7081	+1.0104	+1.0604	+1.1566	+1.1785

New Z Matrix of Gents Drudgery
(Case III Model)

	1	2	3	4	5	6	7	8	9	10
1.	—	.711	2.003	1.142	.832	1.740	1.465	0.238	.832	.832
2.	-.859	—	.280	.976	.280	.656	1.519	1.124	.976	.656
3.	-.463	-.055	—	.192	.192	.055	.164	.222	.164	.463
4.	-.701	-.511	-.511	—	.436	.436	.701	.795	.343	.343
5.	-.278	-.080	-.278	-.238	—	-.670	.382	.238	.151	.382
6.	-.578	-.186	-.079	-.236	-.665	—	.318	.079	.186	.150
7.	-.409	-.362	-.198	-.319	-.319	.267	—	.267	.486	.028
8.	-.253	-1.016	-1.016	-1.373	-.754	-.253	-1.016	—	.593	.217
9.	-1.040	-1.040	-.889	-.699	-.563	-.699	-2.175	-.699	—	.429
10.	-.910	-.612	-2.192	-.612	-1.250	-.493	-.111	-.224	-.376	—
ΣZ_{ij}	-5.491	-3.151	-1.858	-1.551	-.481	+.505	+1.247	+2.140	+3.355	+3.500
$\Sigma Z_{ij}/n$	-.5491	-.3151	-.1858	-.1551	-.0481	+0.0505	+0.1247	+0.2140	+0.3355	+0.3500
Add highest value to each cell	-.5491 +	-.5491 +	-.5491 +	-.5491 +	-.5491 +	-.5491 +	-.5491 +	-.5491 +	-.5491 +	-.5491 +
Scale values	0	+0.2340	+0.3633	+0.3940	+0.5010	+0.5996	+0.6738	+0.7631	+0.8846	+0.8991

Fij Matrix of Pooled Data

	1	2	3	4	5	6	7	8	9	10
1.	—	43	39	39	42	43	52	54	47	43
2.	17	—	43	55	56	47	49	47	38	49
3.	21	17	—	45	48	49	48	55	47	40
4.	21	5	15	—	36	56	52	37	34	42
5.	18	4	12	24	—	48	49	41	46	32
6.	17	13	11	4	12	—	50	47	43	51
7.	8	11	12	8	11	10	—	48	42	42
8.	6	13	5	23	19	13	12	—	44	36
9.	13	22	13	26	14	17	8	16	—	27
10.	17	11	20	8	8	9	8	23	23	—

P_{ij} Matrix of Pooled Data

	1	2	3	4	5	6	7	8	9	10
1.	—	.71	.65	.65	.70	.91	.86	.90	.78	.71
2.	.29	—	.71	.91	.93	.78	.81	.78	.63	.81
3.	.35	.29	—	.75	.80	.81	.80	.91	.78	.66
4.	.35	.09	.25	—	.60	.93	.86	.61	.56	.70
5.	.30	.07	.20	.40	—	.80	.81	.68	.76	.86
6.	.09	.22	.19	.07	.20	—	.83	.78	.71	.85
7.	.14	.19	.20	.14	.19	.17	—	.80	.70	.70
8.	.10	.22	.09	.39	.32	.22	.20	—	.73	.60
9.	.22	.37	.22	.44	.24	.29	.30	.27	—	.61
10.	.29	.19	.34	.30	.14	.30	.30	.40	.39	—

Z_{ij} Matrix of Pooled Data (Case V Model)

	1	2	3	4	5	6	7	8	9	10
1.	—	.553	.553	1.341	1.476	.772	.332	.878	.878	.772
2.	-.553	—	.385	.385	.524	.553	.772	1.080	.553	1.282
3.	-.553	-.385	—	.764	.842	.878	.772	.842	.412	1.341
4.	-1.341	-.385	-.764	—	.253	1.476	.151	1.080	.524	.279
5.	-1.476	-.524	-.842	-.253	—	.842	.706	.878	1.080	.468
6.	-.772	-.553	-.878	-1.476	-.842	—	.553	.954	1.036	.722
7.	-.332	-.772	-.772	-.151	-.706	-.553	—	.524	.279	.613
8.	-.878	-1.080	-.842	-1.080	-.878	-.954	-.524	—	.524	.842
9.	-.878	-.553	-.412	-.524	-1.080	-1.036	-.279	-.524	—	.253
10.	-.772	-1.282	-1.341	-.279	-.468	-.722	-.613	-.842	-.253	—
ΣZ_{ij}	-7.555	-4.981	-4.823	-1.363	-.879	+1.256	+2.723	+4.346	+5.135	+6.572
$\Sigma Z_{ij}/n$	-.7555	-.4981	-.4823	-.1363	-.0879	+.1256	+.2723	+.4346	+.5135	+.6572
Add highest negative value to each cell	-.7555 -	-.7555 +	-.7555 +	-.7555 +	-.7555 +	-.7555 +	-.7555 +	-.7555 +	-.7555 +	-.7555 +
Scale values	0	+.2574	+.2732	+.6192	+.6676	+.8811	+1.0278	+1.1901	+1.2690	+1.4127

New Z Matrix of Pooled Data
(Case III Model)

	1	2	3	4	5	6	7	8	9	10
1.	—	1.126	.784	1.126	1.067	.784	2.199	1.572	2.611	1.126
2.	-.474	—	.474	.066	1.266	1.151	.753	.284	.662	.753
3.	-.314	-.451	—	.715	.686	.549	.686	.629	1.093	.336
4.	-.124	-.174	-.197	—	-.189	-.332	.215	.124	.174	.233
5.	-.156	-.440	-.251	.251	—	-.075	.262	.210	.139	.322
6.	-.135	-.471	-.236	.518	.089	—	.379	.053	.098	.184
7.	-.476	-.387	-.378	-.421	-.387	-.467	—	.231	.371	.231
8.	-1.624	-.698	-1.623	-1.163	-1.485	-.317	-1.102	—	-1.289	.587
9.	-.929	-.560	-.972	-.560	-.339	-.202	-.610	+.444	—	.183
10.	-1.187	-1.884	-.884	-2.223	-2.318	-1.125	-1.125	-.599	-.543	—
ΣZ_{ij}	-5.419	-3.939	-3.276	-1.095	-0.610	-0.043	+1.657	+2.948	+3.316	+3.955
$\Sigma Z_{ij}/n$	-.5419	-.3939	-.3276	-.1095	-.0610	-.0043	+.1657	+.2948	+.3316	+.3955
Add highest negative value to each cell	-.5419	-.5419	-.5419	-.5419	-.5419	-.5419	-.5419	-.5419	-.5419	-.5419
Scale values	0	+.1480	+.2143	+.4324	+.4809	+.5376	+.7076	+.8367	+.8735	+.9374

BIBLIOGRAPHY

Acharya, S. and Patkar, P., Technological infusion and employment condition of women in rice cultivation areas. Paper presented in the conference on *Women in rice farming* held at IRRI, Los Banos, Philippines, 1983.

Aggarwal, B., Rural women and the high yielding rice technology in India. Paper presented in the conference on *Women in rice farming* held at IRRI, Los Banos, Philippines, 1983.

Ahuja, Krishna., Women Technology and Development process. Economic and Political Weekly. 14(36), 1979.

Anandalakshmy, S. and Kelkar, G., *Role of women in post-harvest food conservation*. United Nations University, Tokyo, 1980.

Anandalakshmy, S. and Sawhney, H.K., Women's contribution to the productive processes in agriculture and the impact of modernisation. Paper presented at the international Conference on *Appropriate Agricultural Technologies for farm Women* held at New Delhi, 1988.

Anonymous, *National Accounts Statistics*. Central Statistical Organisation, Government of India, New Delhi, 1987.

Apte, M., Rural women and development in India. *Indian Journal of Social Work*, 39(4), 1979.

Arya, B.S., Role of family in decision-making as related to farm operations : A study conducted in Kanjhawala block, Delhi. M.Sc. Thesis. Indian Agricultural Research Institute, New Delhi., 1964.

Awasthy, I. *Rural Women of India - A Socio-economic Profile of Jammu Women*. B.R. Publishing Corporation, Delhi, 1982.

Bajwa, Anita, Decision-making by rural women in selected household activities. M.Sc. Thesis. Punjab Agricultural University, Ludhiana, 1984.

Bajwa, Gurdeep K., A comparative study of role performance of rural women from farm owning and non-farm owning families in selected agricultural operations. M.Sc. Thesis. Punjab Agricultural University, Ludhiana, 1983.

Banerjee, Jayanti, Implications of technology for women in the Rural Sector. In: *Women and Technology*. Jain, S.C. (ed.). Rawat Publications, Jaipur, pp. 7-16, 1985.

Bates, J., A model for the science of decision. Philosophy of Science. 21 (326), 1954.

Beal, G.M. and Sibley, D.N., Adoption of agricultural technology by the Indian Guatemala. *Rural Sociology report No. 62*. Department of Sociology and Anthropology, Iowa State University, Ames, Iowa, 1967.

Bhatt, K.P., A leading role in agriculture and animal husbandry. *Indian Farming*, 25(39), 1975.

Bloom, B.S., Engelhardt, M., Furst, E., Hill, W. and Krathwal, D.R., *Taxonomy of educational objectives - The cognitive domain*. Orient Longmans, New York, 1956.

Boserup, E., Towards models of development. In: Welsley Editorial Committee: *Women and National Development*. The University of Chicago Press, Chicago, 1977.

Castillo, C.T., The changing role of women in rural societies: A summary of trends and issues, *Seminar Report No. 12*, Agricultural Development Council, New York, 1977.

Census of India, Provisional population totals, workers and non-workers. Series-I, Paper-3, India. pp. 31-59, 1981.

Chakravarty, S., Farm women labour: Waste and exploitation. *Social Change*, 5(9), 1975.

Chapin, F.S., A quantitative scale for ranking the home and social environment of middle class families: A first approximation of the measurement of socio-economic status. *Journal of Educational psychology*, 19(99), 1928.

Chatterjee, A., A statistical profile - Landless agricultural women worker. *Indian Farming*, 25(31), 1975.

Chaudhry, C., A study on the participation and interest of rural school going adolescent farm girls in food production and home management. M.Sc. Thesis. Indian Agricultural Research Institute, New Delhi, 1973.

Choudhary, T.P.S. and Snarma, B.M., Female labour of the farm family in agriculture. *Agricultural Situation in India*, 16 (643), 1961.

Craven, Ruby, The family role in decision-making. *Journal of Cooperative Extension*, 1 (23), 1963.

Dahama, O.P. *Extension and rural welfare*. Ram Prasad and Sons, Agra, p. 39, 1976.

Dak, T.M., Sharma, M.L. and Jain, Rajni, Role of women in agriculture - An investigation into socio-economic correlation affecting changes in their work participation. Paper presented at the seminar on 'Role of development programmes on socio-economic status of women', held at Haryana Agricultural University, Hisar, 1986.

Deepali, M.N., A study on the knowledge and participation of rural women in agricultural operations with respect to paddy crop and their value orientation in Dharwad district. M.Sc. Thesis. University of Agricultural Sciences, Bangalore, 1979.

Devadas, R.P., Thangamanik, S. Mithu, Role of selected farm women in agricultural operations. *Indian Journal of Home Science*, 6 (49), 1972.

Devi, L.A. and Reddy, S.V., Role expectation and role performance of rural women in farm activities. *Indian Journal of Extension Education*, 20 (27), 1984.

Dey, Jenny, Women in Rice Farming System Network - Country profile on India. IRRI, Philippines, 1985.

Dhar, A., Rural women of India. *Kurukshetra*, 26(4), 1978.

Dhiman, P.C. and Khirwar, A.S., White revolution for welfare of rural women. Paper presented in the seminar on 'Role of Development programme on socio-economic status of women' held at Haryana Agricultural University, Hisar, 1986.

Dixon, R.B., Women in Agriculture : Counting the Labour Force in Developing Countries. *Population and Development Review*, 8(539), 1982.

Edward, A.L., *Techniques of attitude scale construction*. Vakils, Feffer and Simons Pvt. Ltd., Bombay, 1969.

English, H.B. and English, A.C., *A comprehensive dictionary of psychological analytical terms*. Green and Co. Longman's New York, 1958.

Gandhi, S. Yadav, L. and Singal, S., Participation of women in agriculture. A Haryana study. Paper presented in a seminar on 'Role of Development programme on socio-economic status of women' held at Haryana Agricultural University, Hisar, 1986.

Ghosh, B. and Mukhopadhyay, S.K., Displacement of the female in the Indian labour force. *Economic and Political Weekly*, 19(47), 1984.

Ghosh, S.S., Role of village women in decision-making process related to farm economic business. M.Sc. Thesis. Bidhan Chandra Krishi Viswa Vidyalaya, Kalyani, 1987.

Gross, I.H. and Crandall, E.W., *Management for modern families*. Appleton-Century-Crofts, New York, pp. 303-304, 1963.

Gulati, L., *Profiles in Female Poverty - A study of Five Poor Working Women in Kerala*. Hindustan Publishing Corporation, New Delhi, 1984.

Hanger, J. and Maris, J., Women and household Economy. In: Robert, C. and John, M. (Eds.), *Mewar — An irrigated settlement in Kenya*, Munich, Weltforum, 1973.

Haque, A., Study of need, interest, aspirations and leisure time activities of school going rural girls. M.Sc. Thesis. Indian Agricultural Research Institute, New Delhi, 1968.

Hira Nand, Sharma, J.P. and Janardanjee, Contribution of farm women to farm operations and decision making thereon with special reference to potato. Central Potato Research Institute, Shimla, 1988.

I.L.O., International labour organisation contribution to the United Nation. *Decade for Women*, P. 9, 1975.

I.L.O., International Labour Organisation Report, 1979.

Jatcly, Surinder, Impact of planned social change and modernisation on rural women - A pilot study in western U.P. *ICSSR Research Abstracts*, 10(1), 1981.

Jhansi Rani, G., Decision-making patterns among rural families and type of participation by rural women in selected areas of decisions. M.Sc. Thesis. Andhra Pradesh Agricultural University, Hyderabad, 1981.

Kabir, K. Chen, M., In: Whyte and Whyte, *op. cit.*, p.29, 1978.

Kakar, S., Feminine identity in India. In: Ghadially, Rehana (ed.) : *Women in Indian Society*. Sage Publications, New Delhi, 1988.

Kanjla, J., A study of knowledge and extent of participation of women in dairy management practices. M.Sc. Thesis. Punjab Agricultural University, Ludhiana, 1980.

Kaur, Malkit, *Female work participation in Rural Haryana: Impact of technological advancement in agriculture*. D.K. Publisher, New Delhi, 1986.

Kaur, Satnam, *Women in Rural Development: A case study*. Mittal Publications, Delhi, 1987.

Khan, M.A. and Ayesha, N., *Status of Rural Women in India*. Uppal Publishing House, New Delhi, 1982.

Laxmi Devi, A., *Rural women: Management in Farm and Home*. Northern Book Centre, New Delhi, 1987.

Lepcha, Tseten Tashi, A study of farm women's contribution to agricultural production in eastern district of Sikkim State. M.Sc. Thesis. Gujarat Agricultural University, Anand, 1987.

Majumdar, D.N., *Caste and Communication in an Indian village*. Asia Publication House, Bombay, 1962.

Majumdar, M., Sectoral plan for rural women for a place in the sun. Indian Express, 1980.

Majumdar, V., Women workers in changing economy. Yojna, 19(7), 1975.

Manohar, K.M., *Socio-Economic Status of Indian Women*. Seema Publication, Delhi, 1983.

Mies, M., *Indian women in subsistence and agricultural labour. World employment programme, Research working paper*. ILO, Geneva, 1984.

Mosteller, F., Remarks on the method of paired comparisons: III. A test of significance for paired comparisons when equal standard deviations and equal correlations are assumed. *Psychometrika*, 16(207), 1951.

Mulay, S., Role of farm women in Green Revolution. *Indian Journal of Adult Education*, 35(90), 1974.

Mumford, I., *The city in History*. Secker and Warburg, London, 1961.

Munjal, S., An exploratory study of rural women's allocation, decision and activity pattern in Karnal district. M.Sc. Thesis. Haryana Agricultural University, Hisar, 1984.

Muse, Marianne, *Kitchen equipment and arrangement*. Agriculture Experiment Station, Vermont, p. 19, 1954.

Nagre, M.L., The role of family members in decision making as related to selected farm operations. M.Sc. Thesis. Marathwada Agricultural University, 1976.

Naine, M., Kherde, R.L. and Lokhande, M.R., Performance of farm women towards food related tasks. *Indian Journal of Extension Education*, 4(35), 1966.

Nancy, H.J. and Bay, E.G., *Women in Africa studies in social and economic changes*. Standard University Press, California, 1970.

Palmer, I., Rural women and the basic needs approach to development. *International Labour Review*, 115(34), 1977.

Pandey, U.K., Malik, H.S., Suhag, K.S. and Rai, L., Participation of rural women in Haryana Agriculture. *Agricultural Situation in India*, pp. 969-974, 1988.

Panikar, P.G.K., *Adoption of high yielding varieties of Rice in Kerala-A study of selected villages in Palghat and Kuttand.* Centre for Development Studies, Trivandrum, 1983.

Parikh, Indira, J. and Garg. P.K., *Indian Women: An inner dialogue.* Sage Publications, New Delhi, 1989.

Pratibha, New *Agricultural Technology and Employment of Women.* M.Sc. Thesis. Lady Irwin College, New Delhi, 1981.

Punia, R.K., Changing Agrarian structure in Haryana. Ph.D. Thesis. Haryana Agricultural University, Hisar, 1980.

Puri, Shashi, A study on decision-making patterns in rural families with special reference to role of women. M.Sc. Thesis. Lady Irwin College, New Delhi, 1968.

Puri, Shashi, Work roles and decision-making pattern of farm wives and husbands. Ph.D. thesis. Indian Agricultural Research Institute, New Delhi, 1972.

Puri, Shashi, *Role of farm women in Animal Husbandry programme.* Kurukshetra/22(9), 1975.

Rai, P., Women labour. *Yojana*, 19(18), 1975.

Rani, Jhansi and Bhave, A.M., Decision making pattern among rural families. *Indian Journal of Adult Education*, 42(17), 1981.

Reick, R.E., An empirical measure of decision-making as an evaluator of farm and home development in wisconsin. Ph.D. Thesis. Cornell University, 1960.

Rogers, E.M. and Havens, A.E., *Diffusion of Innovations.* The Free press of Glencoe, New York. pp. 81-86, 1967.

Ruth, S., *Issues in Feminism.* Houghton Mifflin Company, Boston., 1980.

Sahay, B.N., *Pragmatism in Development: Application of anthropology.* Bookhive Publishers and Booksellers, New Delhi, p.89, 1969

Saikia, A., Seasonal variation of female participation on farm. *Kurukshetra*, 31(15), 1983.

Saikia, P.D. and Gogoi, S.K., Women labour in North-East India. *Yojana.* 25(11), 1981.

Saini, H., Time task analysis of farming and non-farming home makers. M.Sc. Thesis. Punjab Agricultural University, Ludhiana, 1983.

Satapathy, C., *Women in agriculture: A study at adopted villages of OUAT*. Research Project Report, Orissa University of Agriculture and Technology, Bhubaneshwar, 1987.

Sawhney, H.K. and Devnani, H.D. Impact of farm mechanisation on farm employment. M.Sc. Thesis. Lady Irwin College, New Delhi, 1976.

Saxena, Daya and Bhatnagar, Suman, R., A comparative study of time utilization pattern of tribal and non-tribal women in Rajasthan. *Indian Journal of Extension Education*, 21 (83), 1985.

Sen, Ilina, Class and gender in work time allocation in Himachal Pradesh (Hoshangabad). *Economic and Political Weekly*, 1988.

Sharma, D.K. and Singh, T.R., Participation of women in decision related to farm business. *Indian Journal of Extension Education*, 6(43), 1970.

Singh, A., *In Indian Society of Agricultural Economics: Problems of Farm Mechanisation*. Seminar-IX, Bombay, 1972.

Singh, T.R., Participation of rural women in agricultural operations in NES block, Jabalpur. M.Sc. Thesis. Jawaharlal Nehru Krishi Vishva Vidyalaya, Jabalpur, 1968.

Sinha, S.D., *Impact of technological development in agriculture on women in rural areas: A two village profile*. ICSSR, New Delhi, 1980.

Sridevi, K., Time management of rural women in home and farm activities. M.Sc. Thesis. Andhra Pradesh Agricultural University, Hyderabad, 1984.

Srivastava, J.C., Harnessing Technology for improving the quality of life in rural women. In: *Women and Technology*. Jain, S.C. (ed.). Rawat Publications, Jaipur, pp. 38-74, 1985.

Statistical Abstract of Haryana, 1984-85.

Stroller, A.M., Class structure and female autonomy in rural jower. In: Wilsely Editorial Committee, op. cit., p-77, 1977.

Sunita, An exploratory study on expectation and performance of women labour engaged in paddy cultivation. M.Sc. Thesis. Haryana Agricultural University, Hisar, 1986.

Swaminathan, M.S., *Imparting a rural women user's perspective to Agricultural Research and Development*. IRRI, Philippines, 1985.

Thangamani, K., Role of selected farm women in agricultural operations. M.Sc. Thesis. Madras University, Madras, 1971.

Thurstone, L.L., A law of comparative judgement. *Psychology Review*, 34(273), 1927.

Trivedi, G., Measurement and analysis of socio-economic status of rural families - Kanjhawala block. Ph.D. Thesis. Indian Agricultural Research Institute, New Delhi, 1963.

United Nations, World conference of the united Nations Decade for women: Equality, Development and Peace Document A/CONF. 94/19. Recommendations relating to women and development emerging from conferences held under the Auspices of the United Nations or the specialised Agencies, Nairobi, Kenya, 1975.

United Nations, The world conference on U.N. Decade for women: Equality, Development and Peace. Held at Nairobi (Kenya), 1985.

Varma, O.S. and Malik, B.S., contribution of rural ladies in Farm Operations. *Indian Journal of Home Science*, 15(57), 1984.

Verma, J.C., Women in rural economy. *Kurukshetra*, 26(7), 1978.

Vijayalakshmi, K., A comparative study of the role of women in agriculture in Coimbatore district of Tamil Nadu and in Guntur district of Andhra Pradesh. M.Sc. Thesis. Madras University, Madras, 1976.

Wadley, Susan, Brothers, Husbands and sometimes sons: Kinsmen in North Indian Ritual. *Eastern Anthropologist*, 29(149) 1976.

Whyte, B., In Whyte, R.O. and Whyte, P., op. cit., p. 29, 1978.

Whyte, R.O. and Whyte, P., Rural Asian Women status and environment, Research notes and discussion paper No. 9. *Institute of South Asian Studies, 1978.*

Wilkening, E.A., Joint decision making in farm families as a function of status and role. *American Sociology Review*. 23(187), 1962.

Yadav, Sarla, Role of women in decision making related to farm business. M.Sc. Thesis. Haryana Agricultural University, Hisar, 1982.

INDEX